Sneaky Uses for Everyday Things

How to Turn a Penny into a Radio,
Make a Flood Alarm with an Aspirin,
Change Milk into Plastic, Extract
Water and Electricity from Thin Air,
Turn On a TV with Your Ring,
and Other Amazing Feats

Cy Tymony

**Andrews McMeel
Publishing**

Kansas City

Sneaky Uses for Everyday Things

ISBN-13: 978-0-7407-3859-3
ISBN-10: 0-7407-3859-3

Library of Congress Control Number: 2003102991

06 07 08 09   MLT   13 12

Book design by Holly Camerlinck

Book composition by Kelly & Company,
Lee's Summit, Missouri

Attention: Schools and Businesses

Andrews McMeel books are available at quantity discounts with
bulk purchase for educational, business, or sales promotional use.
For information, please write to: Special Sales Department,
Andrews McMeel Publishing, 4520 Main Street,
Kansas City, Missouri 64111.

Sneaky Uses for Everyday Things

Disclaimer

This book is for the entertainment and edification of its readers. While reasonable care has been exercised with respect to its accuracy, the publisher and the author assume no responsibility for errors or omissions in its content. Nor do we assume liability for any damages resulting from use of the information presented here.

This book contains references to electrical safety that *must* be observed. *Do not use AC power for any projects listed.* Do not place or store magnets near such magnetically sensitive media as videotapes, audiotapes, or computer disks.

Disparities in materials and design methods and the application of components may cause your results to vary from those shown here. The publisher and the author disclaim any liability for injury that may result from the use, proper or improper, of the information contained in this book. We do not guarantee that the information contained herein is complete, safe, or acurate, nor should it be considered a substitute for your good judgment and common sense.

Nothing in this book should be construed or interpreted to infringe on the rights of other persons or to violate criminal statutes. We urge you to obey all laws and respect all rights, including property rights, of others.

Contents

Part I
Sneaky Tricks and Gimmicks

Part II
Sneaky Gadgets and Gizmos

Part III

Security Gadgets and Gizmos

Part IV
Sneaky Survival Techniques

Acknowledgments

Special thanks go to my agent, Sheree Bykofsky, for her enthusiastic encouragement and for believing in this book from the start. I am also appreciative of the assistance provided by Janet Rosen and Megan Buckley at her agency.

I wish to thank Jennifer Fox, my editor at Andrews McMeel, and copy editor, Janet Baker, for their invaluable work.

A warm thank-you goes to Bill Melzer for insights and opinions that helped shape this book.

I am also grateful for the project evaluation assistance provided by Jerry Anderson, Isaac English, Carlos Daza, Sybil Smith, and Serrenity Smith.

And I hope the following is adequate to show my invaluable appreciation and love for Cloise Shaw. Thanks, Mom. I love you.

Introduction

"Life . . . is what we make it."
—William James

You don't have to be 007 to adapt unique gadgets, secure a room from intruders, or get the upper hand over aggressors. Anyone can learn how to become a real-life MacGyver in minutes, using nothing but a few hodgepodge items fate has put at our disposal. Sometimes you have to be sneaky.

Sure, it never hurts to have the smarts of Einstein or the strength of Superman, but they're not necessary with *Sneaky Uses for Everyday Things*. When life puts us in a bind, the best solution is frequently not the obvious one. It'll be the sneaky one.

Solutions to a dilemma can come from the most unlikely sources:

- A motorist stranded with a bad heater-valve gasket made a new one by cutting and shaping the tongue from an old track shoe. It worked well enough to get him home safely.

- U.S. prisoners of war devised a stealthy makeshift radio receiver using nothing more than a razor blade, a pencil, and the wire fence of the prison camp as an antenna.

- Convicts at Wisconsin's Green Bay Correctional Institution scaled the prison walls using rope they braided from thousands of yards of dental floss.

- On September 11, 2001, a window washer trapped in a Twin Towers elevator with five other passengers used his squeegee to pry open the doors and chisel through the wall to escape the inferno.

People rarely think about the common items and devices they use in everyday life. They think even less about adapting them to perform other functions. For lovers of self-reliance and gadgetry, *Sneaky Uses for Everyday Things* is an amazing assortment of more than forty fabulous build-it-yourself projects, security procedures, self-defense and survival strategies, unique gift ideas, and more.

Did you know that the coins in your pocket can generate electricity or receive radio signals? Want to know what household item can identify counterfeit paper currency? How to turn milk into plastic or glue? How to locate directions using the sun or the stars? How to make a compass without a magnet, extract water from thin air, use water to start a fire, or make a ring that can turn on your TV? It's all here. Even wire hangers and coffee-cream-container tops get their moment in the sun.

Sneaky Uses for Everyday Things does not include conventional projects found in most crafts and household hints books. Nor are instructions supplied for first aid, fishing, making a shelter, or spotting edible plants. The Resources section at the back includes lists of books and Web sites for obtaining science tricks, frugal facts, and camping information.

Sneaky Uses for Everyday Things avoids projects or procedures that require expensive or unusual materials not found in the average home. No special knowledge or tools are needed. Whether you like to conserve resources or like the idea of getting something for nothing, you can use the book as a practical tool, a fantasy escape, or a trivia guide; it's up to you. "Things" will never appear the same again.

Let's start now. You can do more than you think!

Part I

Sneaky Tricks and Gimmicks

You too can do more with less! Many household items you use every day can perform other functions. Using nothing but a few supplies like paper clips, rubber bands, and refrigerator magnets, you can quickly make unique gadgets and gifts.

Want to know how to tell real paper currency from fake? How to make plastic and glue out of milk? Generate electricity from fruits?

If you have an insatiable curiosity for sneaky secrets of everyday things, look no further. The projects that follow can be made in no time. Start your entry into clever resourcefulness here.

The Fear of Small Sums:
Detect Counterfeit Bills

Whether it's a hundred-dollar bill or a one, getting stuck with counterfeit money is a fear many of us have. In the United States in 2002, $43 million in fake currency was circulated. When counterfeit currency is seized, neither consumers nor companies are compensated for the loss. So what can we do about it? This project describes two methods to tell good currency from bad.

The first method is a careful visual inspection of the bill. Compare a suspect note with a genuine note of the same denomination and series. Look for the following telltale signs:

1. The paper on a genuine bill has tiny red and blue fibers embedded in it. Counterfeit bills may have a few red and blue lines on them but they are printed on the surface and are not really embedded in the paper.
2. The portrait and the sawtooth points on Federal Reserve and Treasury seals are distinct and clear on the real thing.
3. The edge lines of the border on a genuine bill are sharp and unbroken.
4. The serial number on a good bill is evenly spaced and printed with the same color as the Treasury seal.

The second way to verify paper currency is to test the ink. How can we do this in a sneaky way, at home or in the office? Easy: by using one important feature of the ink used on U.S. currency. A legitimate bill has iron particles in the ink that are attracted to a strong magnet. To verify a bill, obtain a very strong magnet or a rare-earth magnet. Rare-earth magnets are extremely strong for their small size. They can be obtained from electronic parts stores and scientific supply outlets. See the Resources section at the back of this book.

You can also use small refrigerator magnets, connecting them end to end to create collectively a much stronger single magnet. See Figure 1.

What's Needed
- Dollar bill
- Strong magnet

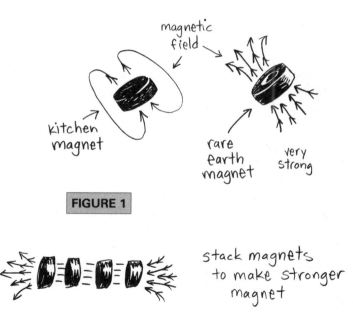

magnetic field

kitchen magnet

rare earth magnet

very strong

FIGURE 1

stack magnets to make stronger magnet

What to Do

Fold the bill in half crosswise and lay it on a table, as shown in
Figure 2. Point the strong magnet near the portrait of the president on the bill, but do not touch it. A legitimate bill will move
toward the magnet, as shown in Figure 3.

Whenever you doubt the authenticity of paper currency, simply
pull out your magnet and perform the magnetic attraction test.
If you create the Power Ring shown in Part II of this book, it can
be used for currency tests too.

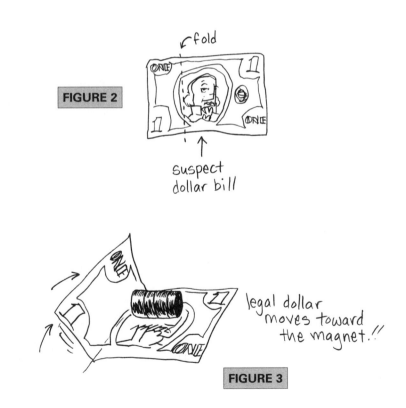

FIGURE 2

fold

suspect
dollar bill

legal dollar
moves toward
the magnet.!!

FIGURE 3

Slushy Fun:
Cheap Gel Packs for Swollen Muscles

People in physically demanding jobs and weekend warriors get muscle aches often. When they do, an icy gel pack can relieve the pain and swelling.

Gel packs work to reduce swelling because they can be fitted around joints to cool them thoroughly. You can save money by making your own version—a slushy pack—from everyday things found in the home.

What's Needed
- Water
- Rubbing alcohol
- Watertight freezer bag—typically 6½" x 5⅞"

water

rubbing alcohol

the kind that seals ↴

waterproof sandwich bag

FIGURE 1

What to Do

Add 1½ cups water and ½ cup alcohol to the plastic bag and seal it. Ensure that the bag is not overfilled. Place bag in the freezer for 3 hours (see Figure 1).

The fluid inside will not freeze solid. Instead, the alcohol keeps the water flexible and slushy for a better fit (see Figure 2).

When needed, remove the bag from the freezer and apply it to the swollen area as shown in Figure 3. To prevent frostbite or cold burns, place a towel or cloth between the plastic and the skin.

Once you're done, place the slushy pack back in the freezer for future use.

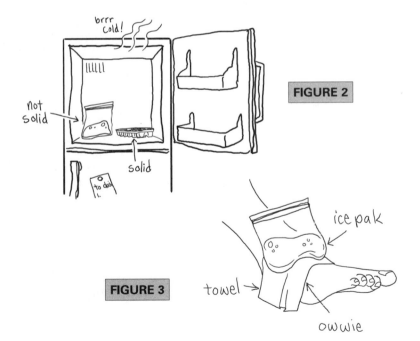

FIGURE 2

FIGURE 3

Got Plastic?
Turn Milk into Sneaky Plastic

Have you ever needed plastic molding material for a repair or a craft project? Perhaps you broke off a plastic piece from a toy or an appliance and need to fill in the unsightly gap. Well, you can transform commonplace items in your kitchen into a flexible compound that will do such repairs and even allow you to paint to match.

Believe it or not, you can make a malleable plastic material from plain household milk and only one other ingredient—vinegar. It's easy.

What's Needed
- Milk
- Small pot
- Spoon
- Vinegar
- Strainer
- Jar
- Paper towels

FIGURE 1

What to Do

Pour an 8-ounce cup of milk in a pot and heat it on a stove. Let it warm but not boil. Add a tablespoon of vinegar and stir the mixture. Soon, clumps of a solid material will form on the surface. Continue stirring (see Figure 1).

Place the strainer on top of the jar and pour the mixture through the strainer. Use the spoon to press the clumps and squeeze out the liquid, which is discarded. Remove the material from the strainer and place it on a paper towel. Dab paper towels on top of the material to absorb excess moisture (see Figure 2).

The solid material that has formed is called casein. It separates from milk when an acid, like vinegar, is added. Casein is used in industry to make glue, paint, and some plastics. You can now form the "sneaky plastic" into shape with a mold or use your hands. Allow the shaped material to dry for 1 to 2 days.

Sneaky plastic has many uses. First, it allows you to recycle spoiled milk that would be discarded anyway. You can make impressions of coins and other small objects. You can shape the plastic into parts to replace items such as broken Walkman belt clips. Or you can make a personalized key ring ornament (see Figure 3).

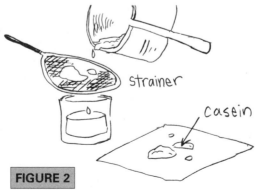

strainer

casein

FIGURE 2

Here are some more ideas for putting this plastic compound to use:

Child-proofing items with sharp edges or points
Toy assembly aid (to hold wood and plastic pieces together)
Caulk for small holes in a boat
Pendant holder
Wheels for carts and toys
Tool handle
Material for a spacer or washer
Guitar pick
Bottle cap
Temporary plumbing repairs
Waterproof container
Fishing lure and float
Replacement button

Sneaky plastic can also be used to create a Power Ring (see Part II under "Superman and Green Lantern Ain't Got Nothin' on Me"). Whether you use the compound for critical repairs or just for fun craft projects, you'll discover it provides plenty of versatility with only a small investment of time.

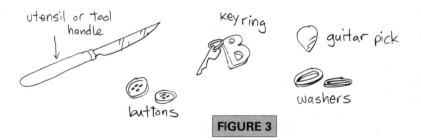

utensil or tool handle

key ring

guitar pick

buttons

washers

FIGURE 3

Need Glue?
Create Sneaky Glue from Milk

If you have an emergency need to secure items together and you're out of glue, don't have a cow. Milk one.

By adding two common ingredients to milk, you can make sneaky glue! When vinegar is added to milk, a sticky substance called casein is formed. By adding baking soda, you can create a gluelike substance.

What's Needed

- Milk
- Small pot
- Spoon
- Vinegar
- Strainer
- Jar
- Paper towels
- Baking soda

FIGURE 1

What to Do

Pour an 8-ounce cup of milk in a pot and heat it on a stove to 250 degrees. Let it warm but not boil. Add a tablespoon of vinegar and stir the mixture. Soon, clumps of a solid material will form on the surface. Continue stirring (see Figure 1).

Place the strainer on top of the jar and pour the mixture through the strainer. Use the spoon to press the clumps and squeeze out the liquid, as shown in Figure 2.

Remove the material from the strainer and place it back in the pot, on the stove. Add ¼ cup of water and a tablespoon of baking soda (see Figure 3). The casein material will begin to bubble. When it stops, use the leftover material as glue.

Note: Wait several hours before using any item secured with sneaky glue.

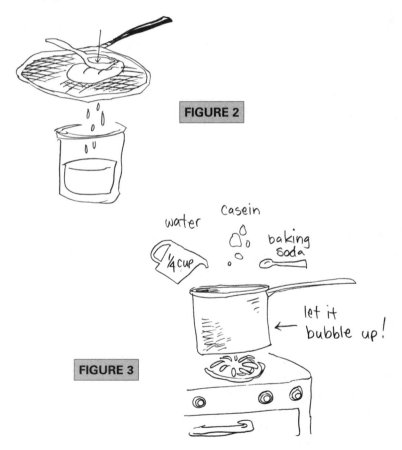

FIGURE 2

FIGURE 3

Spin Thrift:
Make a Videotape Rewinder

Do you want to exercise and lose weight from the comfort of your easy chair? Would you like to do it while you watch television and prevent needless wear and tear of your VCR at the same time? Well, now you can!

If this sounds like a TV-infomercial pitch for a new exercise system, guess again. You'll be able to do all of the above and not spend a dime. If you own a VCR, you know what an inconvenience it is to rewind your videotapes. You must sit and wait for up to five minutes while your VCR is generating lots of heat and wearing out its motor. If you have a rental tape, it must be rewound before it is returned. Some video rental stores will charge a fine if their tapes are not returned fully rewound.

You could purchase an electric videotape rewinder, but they are costly, use electricity, and eventually break down. Some of the cheaper models have even been know to break tapes because they do not properly sense where they end. If you own a camcorder, you know how valuable your limited battery life is. Unnecessarily fast-forwarding and rewinding tapes will quickly drain a camcorder's battery. By using a free portable manual tape rewinder, you will save your battery power for recording additional scenes in the field.

You can easily make a manual tape rewinder yourself within ten minutes from ordinary household items: a wire garment hanger and a paper clip. Then you can rewind your videotapes while you watch television so it won't seem tedious. I do it all the time.

What's Needed
- Wire hanger
- Paper clip
- Tape (any type)
- Pliers

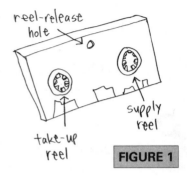

reel-release hole

take-up reel

supply reel

FIGURE 1

What to Do
First, let's find out why you can't just rewind a videotape with a pen or your fingers. Figure 1 illustrates the bottom view of a typical VHS cassette. There is a supply reel, which has the tape wrapped around it, and a take-up reel, which holds the tape after it plays to the end. If you try to turn either reel, they will not move. There is a catch mechanism in the tape shell that prevents tape reel rotation unless a small thin object is pushed through the small reel-release hole.

VCR manufacturers designed their machines this way to prevent damage to a tape in the event of a malfunction. When a tape is loaded in a VCR, a small thin part of the chassis protrudes through the tape shell's reel-release hole, allowing the reels to rotate. In case of a problem, the machine immediately stops the tape and shuts down or ejects the tape to prevent breakage. You can design your rewinding device for a VHS, Betamax, or 8-millimeter cassette tape.

Using pliers, untwist a standard wire garment hanger into a long straight line and cut it in half with the pliers. If you cannot cut the wire, bend it back and forth rapidly, and the heat generated by the motion will eventually cause it to break in two. Then bend a paper clip into the shape shown in Figure 2.

Next, bend the hanger with the pliers into the shape shown in Figure 2. The only size that is crucial is the width of the **C**-shaped end. If you want to rewind a VHS tape, the **C**-shaped end should be approximately ⅗ inch wide.

For a Beta tape, the **C** shape width should be approximately ⁷⁄₁₀ inch. An 8-millimeter tape rewinder requires a ⅖ inch **C** shape at the end of the hanger. These dimensions are approximate because you'll need to bend the **C**-shaped end outward slightly to attain a snug fit when it's positioned in the cassette's tape reel sprocket.

Note: If you wrap the **C** area of the hanger with tape—masking tape, duct tape, or electrical tape will do nicely—it will reduce the chances that the videotape rewinder will slip out of the sprocket reel.

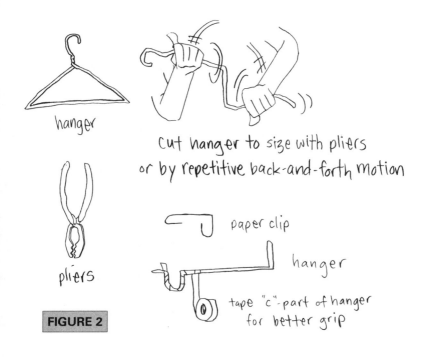

hanger

Cut hanger to size with pliers
or by repetitive back-and-forth motion

pliers

paper clip

hanger

tape "c"-part of hanger
for better grip

FIGURE 2

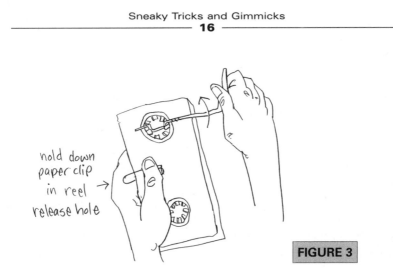

hold down
paper clip
in reel →
release hole

FIGURE 3

Now push the paper clip into the release hole of the cassette. See Figure 3 for the location of the release hole on Beta and VHS tapes (8-millimeter tapes have no reel-release mechanism).

Press the **C**-shaped end of the rewinder into the take-up reel sprocket of the cassette. Hold the paper clip in the release hole in one hand and turn the rewinder's crank arm with the other (see Figure 4). If the rewinder slips out of the reel sprocket, bend the **C** shape outward to achieve a tighter fit.

There you have it: a no-cost manual videotape rewinder!

BONUS APPLICATION:
MINI SPIN ME, AN AUDIOCASSETTE REWINDER

You've seen how easy it is to make a videotape rewinder. Why not make one for your audiocassettes too? If you use a portable cassette recorder or a Walkman tape player, you can extend their battery life tremendously by rewinding the tapes yourself.

What's Needed

- One large paper clip

What to Do

Bend the paper clip into the shape shown in Figure 1. The **C**-shaped end should be approximately ⅜ inch wide. The **C**-shaped end should fit tightly enough to produce a *click* sound when inserted into the cassette's reel sprocket.

Press the **C** end into the tape reel sprocket and turn the rewinder's crank handle (see Figure 2).

Look through the cassette's tape window to see when the tape reaches the end (or desired position). It only takes about three minutes to rewind a 60-minute cassette from one end to the other.

Once you've finished, you can feel confident that you've saved a little battery power and burned calories to boot.

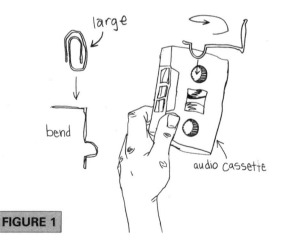

FIGURE 1

Getting Wired:
Sneaky Wire Sources Are Everywhere

It will soon become obvious that many projects in this book use electrical wire. In an emergency, you can obtain wire—or items that can be used as wire—from some very unlikely sources.

To test an item's conductivity (its ability to let electricity flow through it), use a flashlight bulb or an LED. (LED is short for light-emitting diode; LEDs are used in most electronic devices and toys as function indicators because they draw very little electrical current, operate with very little heat, and have no filament to burn out.

Lay a small 3-volt watch battery on the item to be tested, as shown in Figure 1. If the bulb LED lights, the item can be used as wire for battery-powered projects.

Ready-to-use wire can be obtained from telephone cord, TV/ VCR cable, headphone wire, earphone wire, and speaker wire, and also from inside toys, radios, and other electrical devices. (*Note:* Some of these sources have from one to six separate wires inside.)

Wire for projects can also be made from take-out food container handles, twist-ties, paper clips, envelope clasps, ballpoint pen springs, fast-food wrappers, and potato-chip bag liners.

You can also use aluminum from the following items:

Margarine wrapper
Ketchup or condiment package
Breath-mint-container label

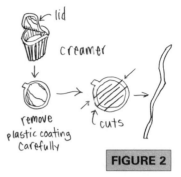

FIGURE 1

FIGURE 2

Chewing-gum wrapper
Trading-card packaging
Coffee-creamer-container lid

You can cut strips of aluminum material from food wrappers easily enough. With smaller items—such as aluminum obtained from a coffee-creamer-container lid—use the sneaky cutting

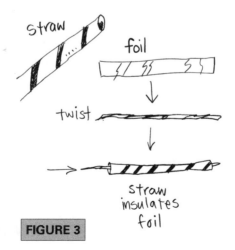

FIGURE 3

pattern shown in Figure 2. Special care must be taken handling the fragile aluminum materials listed. In some instances, aluminum material will be covered by a wax or plastic coating that you may be able to remove.

Note: Wire from aluminum sources is only to be used for low-voltage battery-powered projects.

Figure 3 illustrates how items as small as a cut-up coffee-creamer-container lid can be insulated from other items using a straw, hollow stirrer, or paper.

The resourceful use of items to make sneaky wire is not only intriguing, it's fun, too.

BONUS APPLICATION:
HOW TO CONNECT THINGS

So far, this project has illustrated how to obtain wires from everyday things. Now you'll learn how to connect the wires to provide consistent performance. A tight connection is crucial in electrical projects; otherwise, faulty and erratic operation may result.

Figure 1 shows a piece of insulated wire. Strip the insulation material away to make a connection to other electrical parts. Remove one to two inches of insulation from the end of the wire; see Figure 2.

To connect the wire to a similarly stripped wire, wrap the stripped ends around each other, as shown in Figure 3's three steps.

When connecting the wire to the end of a stiff lead (like the end of an LED), wrap the wire around the lead and bend the lead back over the wire; see Figure 4.

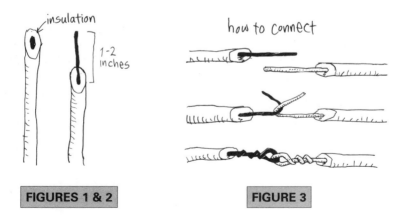

| FIGURES 1 & 2 | FIGURE 3 |

To connect wire to the end of a small battery, bend the wire into a circular shape, place it on the battery terminal, and wrap the connection tightly with tape, as shown in Figure 5.

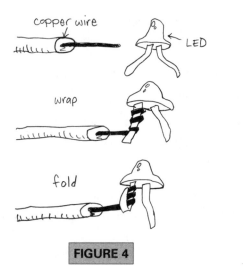

FIGURE 4

FIGURE 5

More Power to You:
Make Batteries from Everyday Things

No one can dispute the usefulness of electricity. But what do you do if you're in a remote area without AC power or batteries? Make sneaky batteries, of course!

In this project, you'll learn how to use fruits, vegetable juices, paper clips, and coins to generate electricity.

What's Needed

- Lemon or other fruit
- Nail
- Heavy copper wire
- Paper clip or twist-tie
- Water
- Salt
- Paper towel
- Pennies and nickels
- Plate

What to Do
THE FRUIT BATTERY

Insert a nail or paper clip into a lemon. Then stick a piece of heavy copper wire into the lemon. Make sure that the wire is close to, but does not touch, the nail (See Figure 1). The nail has become the battery's negative electrode and the copper wire is the positive electrode. The lemon juice, which is acidic, acts as

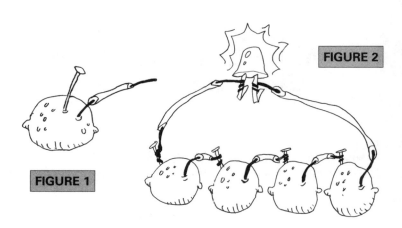

FIGURE 2

FIGURE 1

the electrolyte. You can use other item pairs besides a paper clip and copper wire, as long as they are made of different metals.

The lemon battery will supply about one-fourth to one-third of a volt of electricity. To use a sneaky battery as the battery to power a small electrical device, like an LED light, you must connect a few of them in series, as shown in Figure 2.

THE COIN BATTERY

With the fruit battery, you stuck the metal into the fruit. You can also make a battery by placing a chemical solution between two coins.

Dissolve 2 tablespoons of salt in a glass of water. This is the electrolyte you will place between two dissimilar metal coins.

Now moisten a piece of paper towel or tissue in the salt water. Put a nickel on a plate and put a small piece of the wet absorbent paper on the nickel. Then place a penny on top of the paper (see Figure 3).

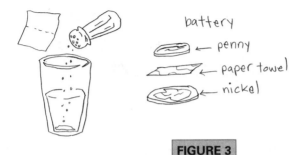

battery

← penny

← paper towel

← nickel

(cut towel
smaller than
nickel!

FIGURE 3

In order for the homemade battery to do useful work, you must make a series of them stacked up as seen in Figure 4. Be sure the paper separators do not touch one another.

The more pairs of coins you add, the higher the voltage output will be. One coin pair should produce about one-third of a volt. With six pairs stacked up, you should be able to power a small flashlight bulb, LED, or other device when the regular batteries have failed. See Figure 5. *Power will last up to two hours.*

Once you know how to make sneaky batteries, you'll never again be totally out of power sources.

Six pairs of
coin batteries
stacked

FIGURE 4

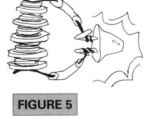

FIGURE 5

You Light Up My Life: Construct Electronic Greeting Cards

To make an impression on someone, give a gift. For a unique and lasting impression, give a handmade sneaky personalized gift. This project will show you how to use parts from discarded toys and gadgets to make electronic greeting cards, posters, and more.

LEDs are found in most electronic devices and in toys and appliances. They are little lights that indicate whether a device or a function is on. Unlike ordinary lightbulbs, these miniature marvels do not have a filament, produce virtually no heat, consume very little power, and (when properly powered) never burn out!

You can obtain LEDs from old discarded toys and other devices. You will need to cut their leads away from the small circuit board, using pliers or wire cutters. You can also purchase LEDs at electronic parts stores locally or from sites shown in the Resources section at the back of the book. You can also obtain blinking LEDs that flash on and off without requiring other components.

What's Needed
- LED
- 3-volt watch battery
- Business card
- Tape
- Wire

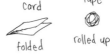
business card
folded

tape
rolled up

insulated wire (with stripped ends)
curled

L E D

insulated wire (with stripped ends)

battery

FIGURE 1

LED

What to Do

Since LEDs require 2 to 5 volts to operate, the best compact power supply for them is a 3-volt lithium watch battery. Since AA, C, and D cell batteries provide just 1½ volts each, you would need two cells to provide 3 volts (although larger cells will last longer).

To see how an LED works, press its leads on both sides of a 3-volt battery, as shown in

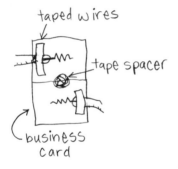

taped wires
tape spacer
business card

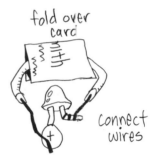

fold over card
connect wires

FIGURE 2

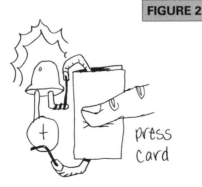

press card

press

you light up my life

Figure 1. If the LED doesn't light, reverse the battery position or reverse the LED leads.

Figure 2 shows how to make a sneaky "touch switch" with a folded business card, stiff wire, and tape. Tape the wires to the card and roll up a piece of tape to act as a spacer. When properly positioned, a slight press of the folded card will connect the wires and light the LED. See the "Invite the Power" section of Part II for methods of controlling LEDs and other devices with a ring.

LEDs can be mounted on greeting cards and bookmarks, belts and bracelets, behind posters, on trophies—and even on clothing! They can add value to items you might otherwise throw away.

Part II

Sneaky Gadgets and Gizmos

All too often people discard older or broken but still functional high-tech gadgets without realizing what other functions they can serve.

Despite the complexity of radios, tape recorders, and other gizmos, this section will illustrate simple, sneaky projects to take advantage of their little-known capabilities.

Want to see how to turn on a TV with your ring, open a room door with a toy car, or adapt a tape recorder into a hearing aid or megaphone? It's here.

Do you know the sneaky method to make a radio out of a penny or how to turn a screw in an FM radio to magically receive aircraft broadcasts? That's here, too, along with bonus applications.

If you're intrigued with high-tech resourcefulness, the following easy-to-build projects will fascinate and delight.

"Superman and Green Lantern Ain't Got Nothin' on Me": **Make a Power Ring**

What can you do with a commonly worn trinket? More than you might think. Here's a novel project you can make with every-day things that will allow you to activate LEDs, toys—even appliances—with your *ring*!

What's Needed

- Toy ring (with a flat surface)
- Small magnet
- Glue

What to Do

The power ring works with a magnet attached to its front surface. It is aimed at a magnetically sensitive switch you will make in the next project.

Glue a small magnet to the surface of a toy ring. You can use a refrigerator magnet or, preferably, a small rare-earth magnet. A rare-earth magnet will allow the power ring to activate devices from a longer distance. Rare-earth magnets can be obtained from electronic part stores and scientific supply outlets (see Figure 1).

You can also create a custom-made ring using the sneaky plastic made from the *Got Plastic?* project in Part I.

mₐgneʈ on ring

FIGURE 1

Power Ring Uses

As a Secret Signaling Device. Attach a small toy compass to a door or window with tape. A friend with a power ring can aim the ring from a secret location and the compass needle will spin, signaling that it's your friend.

To Verify Currency. Fold a bill in half and lay it on a table. Point the power ring close to the portrait of the president on the bill. A legitimate bill has iron particles in the ink and will move toward the magnet.

For an Emergency Compass. Need to make a compass? Use your power ring. Obtain a small thin piece of metal (but not aluminum) and stroke the power ring repeatedly in one direction at least fifty times. You can use a straightened paper clip, a needle, a twist-tie, or a staple from a magazine. Place the metal on a piece of paper or leaf and let it float on the surface of a

FIGURE 2

container of water. The metal will be magnetized enough to be attracted to the Earth's north and south magnetic poles and act as a "Sneaky" compass.

To Activate Toys and Devices. The next project—"Invite the Power"—illustrates how to make a magnetically-activated switch using aluminum foil and a paper clip. With it, only you will be able to activate small lights and buzzers placed inside greeting cards, toys, and gifts as shown in Figure 2.

To Control Appliances. "Invite the Power!" also illustrates how the power ring can be used safely to control household lights and appliances with a device called the X-10® Universal Interface, available from home supply, hardware, and electronic parts stores.

BONUS APPLICATIONS

If you prefer not to activate devices with a ring, a magnet can be mounted in other ways.

Glue a magnet to the end of a plunger dowel (or decorative stick), and you've got a magic wand. Or place a thin rare-earth magnet between two business cards and tape them together. Now you can activate selected devices in the style of a magnetic-strip security card; see Figure 3.

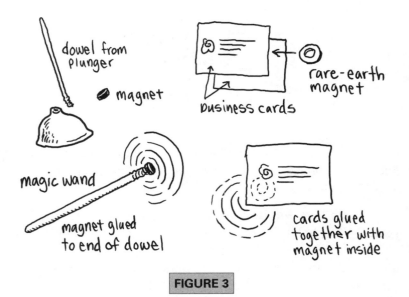

dowel from plunger

magnet

rare-earth magnet

business cards

magic wand

magnet glued to end of dowel

cards glued together with magnet inside

FIGURE 3

"Invite the Power!":
Make Power-Ring-Activated Gadgets

What's the good of having a unique power ring without something for it to control? Not much. After you've made a power ring, as just described, you can use it to activate a variety of devices. This project will show how to make a magnetically sensitive switch. With it, the power ring can turn on a variety of devices, including AC-operated appliances.

What's Needed

- Power ring
- Paper clip or twist-tie
- Aluminum foil
- Business card or cardboard
- LED or flashlight bulb
- Wire
- Tape
- Optional X-10® Universal Powerflash Interface and Appliance Module

power ring

LED

paper twist-
clip or tie

business card

aluminum foil

insulated wire (with stripped ends)

tape

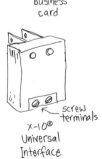

screw terminals

X-10® Universal Interface

button

X-10® Appliance Module

What to Do

To create a magnetically activated switch, tape the aluminum foil to the business card. Then tape a piece of wire to the aluminum foil. Roll up a small piece of tape and place it adjacent to the foil, as shown in Figure 1.

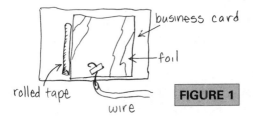

business card

foil

rolled tape

wire

FIGURE 1

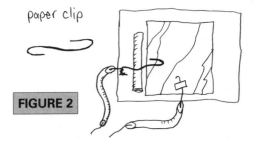

paper clip

FIGURE 2

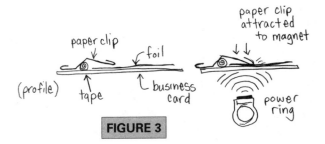

paper clip
attracted
to magnet

paper clip

foil

(profile)

tape

business
card

power
ring

FIGURE 3

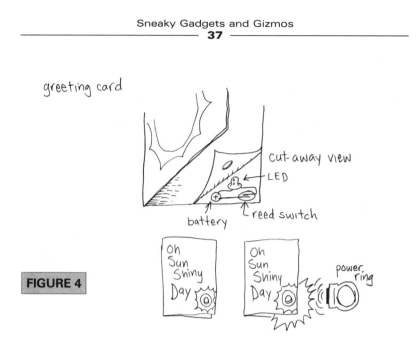

greeting card

cut-away view

LED

battery

reed switch

Oh Sun Shiny Day

Oh Sun Shiny Day

power ring

FIGURE 4

Next, bend the paper clip into an **S** shape and wrap another piece of wire around it. Tape the paper clip near the foil and over the tape so that one end of it rests slightly above, but not touching, the foil; see Figure 2.

Test the switch by aiming the power ring close to the switch. The paper clip should move forward and make contact with the aluminum when the power ring is close; see Figure 3.

Now the ends of the two wires can be attached to a battery and an LED, and when the power ring is aimed at the paper clip the LED will light. Cover the parts with a piece of cardboard and tape the covers closed for protection.

The power-ring–activated switch can be used in a variety of applications, such as activation of LEDs in greeting cards, posters. and jewelry; see Figure 4.

CONTROLLING APPLIANCES WITH A POWER RING

In many homes, people use X-10® controllers and appliance modules to activate TVs, stereos, and lights by remote control. Typically, appliances are plugged into X-10® modules. The controller can turn the appliances on or off from another room, as shown in Figure 5.

There is a special X-10® controller model called the Universal Powerflash Interface that includes two connections for wires to be attached to it. Figure 6 shows how to attach the power-ring–activated switch wires to the Universal Interface contact screws. If you plug in an appliance to an X-10® appliance module, you can turn it on when you aim the ring at the switch; see Figure 7.

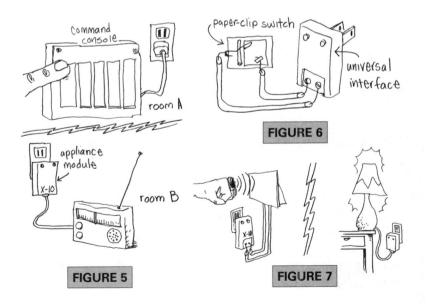

FIGURE 5

FIGURE 6

FIGURE 7

A sneaky switch can be hidden behind or inside of items. Only you will know where to point the power ring and turn on the appliance of your choice.

For more information about X-10® modules and controllers, see www.X10.com.

Gifts of a Feather
You Make Together:
Build Togetherness Gifts

Here's a novel project, which illustrates that some gifts, like people, need each other. The gift set will illuminate an LED only when they are facing each other.

Togetherness gifts work with the same parts and design used in the power ring project, but each gift has a magnet inside that will activate the other gift's magnetically sensitive sneaky switch.

What's Needed

- Two small strong magnets
- Paper clips
- Aluminum foil
- Cardboard
- Two gift boxes
- Two LEDs
- Wire

small magnet

Cardboard

LED

aluminum foil

paper clip

2 gift boxes

insulated wire
(with stripped ends)

What to Do

First, using the instructions given in the "Invite the Power!" section, build two sets of sneaky switches and connect a battery and LED to each as shown in Figure 1.

Mount the sneaky-switch circuits with batteries and LEDs in the two gift boxes. Cut a small hole through which the LEDs can be seen. Then, with the gift boxes facing each other, tape the two magnets so they are positioned opposite the sneaky switches; see Figure 2.

Test the togetherness gift set by placing them close to each other and see if the LEDs will light. If they do not, adjust the positioning of the two magnets until they do.

Last, secure the covers on the gift boxes with tape and write a passage on the outside (or tape a card to each box) with an appropriate message: For instance, *We both need each other.*

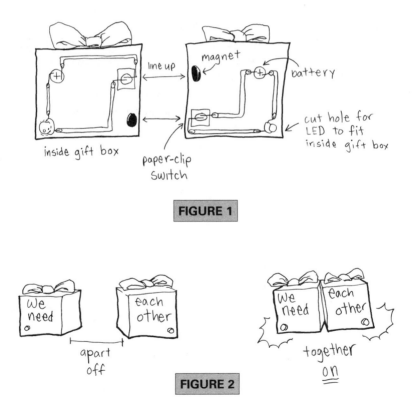

FIGURE 1

FIGURE 2

Listen Impossible:
Make Recordings Only You Can Hear

Many people are reluctant to preserve their personal thoughts on tape out of fear of someone's finding and listening to it later. If that sounds like you, this project will illustrate a sneaky technique you can use to make private messages that stay private.

How is a tape recorded? As seen in Figure 1, a cassette tape is positioned in a recorder's case so that the tape moves past the RECORD/PLAY tape head at a 90-degree angle. This is called the tape head's "azimuth angle." The tape-head position can be adjusted by turning a small screw. If this is done, the quality of a prerecorded tape will suffer.

However, if a blank tape is recorded on a machine with a repositioned tape head, the quality will be fine—but only on that tape recorder. If the tape is played on another recorder, the sound will be garbled.

You can take advantage of azimuth loss by recording personal messages on your tape recorder with the tape head's position changed and then readjust the head so you can use the recorder in a normal fashion. If someone plays the tape, the signal will be virtually inaudible. When needed, you can play back the tape after adjusting the tape head to hear the sounds clearly.

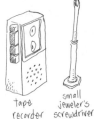

Phillips head

tape recorder

small jeweler's screwdriver

What's Needed
- Tape recorder
- Small jeweler's screwdriver

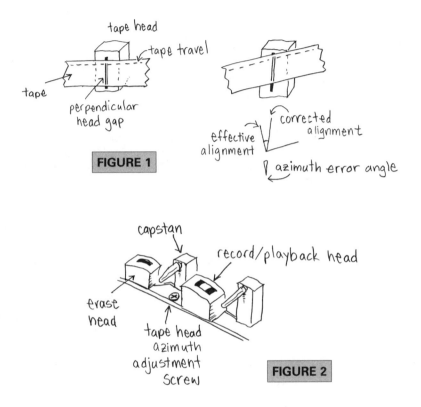

tape head

tape travel

tape

perpendicular
head gap

FIGURE 1

corrected
alignment

effective
alignment

azimuth error angle

capstan

record/playback head

erase
head

tape head
azimuth
adjustment
screw

FIGURE 2

What to Do

Since the screw slot that allows adjustment of the tape-head
position is cross-shaped and very small, you will need a tiny
Phillips screwdriver. This tool is commonly found in eyeglass
repair kits available from drugstores.

Place a tape in a portable tape recorder and press the PLAY
button. Look carefully at the tiny hole or slot on the case near
the tape head. As shown in Figure 2, there will be a small screw
that allows you to adjust the head position. By turning the screw

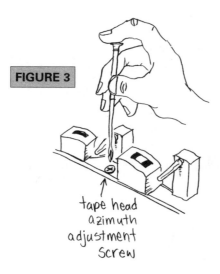

FIGURE 3

tape head
azimuth
adjustment
screw

in either direction, you can alter the azimuth angle of the tape head (see Figure 3).

Note the original position of the screw. Then, with a small screwdriver, adjust the tape head while playing a prerecorded tape. You'll hear the sound quality decrease as you turn the screw in either direction from its starting point. Leave the screw in the position that produces the lowest audio and quality level.

Next, place a blank tape in the recorder and, with some music playing in the background (to make it more difficult for an eavesdropper to understand what is being said), press the RECORD button and speak into the microphone. When you play back the tape, you'll hear your voice clearly. But if you play the tape in another tape player, your voice will be incomprehensible. When you want to hear a specially recorded tape, simply adjust the tape head to where it was originally. Now you can make personal tapes wherever you wish, and only you can successfully play them back.

For Your Ears Only:
Use a Tape Recorder as a Sound Amplifier

In the movies, spies always seem to have a sound-snooping device so they can hear across a room or through walls. But you don't have to request special ordnance from Q to have this sneaky accessory. With a common tape recorder, an earphone, and a small microphone, you'll be able to make a sneaky hearing aid, stethoscope, and more.

Sneaky Fact: A tape recorder sends sound signals to its earphone jack while it is in the RECORD mode. You can prove this by listening with an earphone while you record a tape.

What's Needed
- Portable tape recorder
- Earphone
- Small microphone
- Suction cup (optional)
- Hanger
- Cardboard
- Aluminum foil

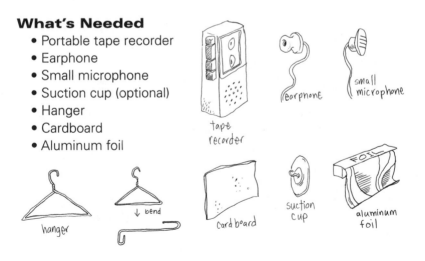

tape recorder

earphone

small microphone

hanger

↓ bend

cardboard

suction cup

aluminum foil

tape recorder

mike

earphone

FIGURE 1

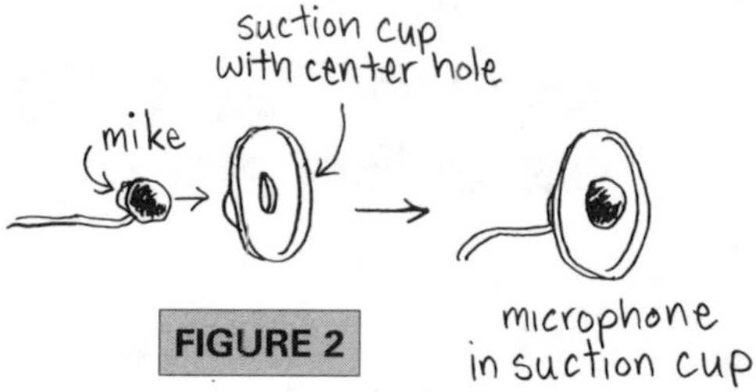

suction cup
with center hole

mike

FIGURE 2

microphone
in suction cup

What to Do

When placed in the RECORD mode, a tape recorder will amplify sounds from the microphone and send signals to the tape head. It also sends audio signals to the earphone jack. In essence, it is an audio amplifier.

To use this function, place a cassette tape in the recorder. Then plug the microphone and earphone into their appropriate jacks. Attach the wires to the plug as shown in Figure 1. Press the RECORD button; then press the PAUSE button. This will stop the motor from turning and will save battery power.

Turn up the volume control. You'll be able to monitor distant sounds with the microphone, and they will be heard from the earphone. Congratulations! You've just made a portable sneaky sound-snooper system.

If you puncture a small hole in a suction cup and place the microphone in it, you can place the microphone on a door or wall and hear sounds through it. Place the microphone on your chest and you'll have an electronic stethoscope (see Figure 2).

BONUS APPLICATION

For sneaky long-distance listening, cut a large sheet of aluminum foil and a piece of cardboard (11 by 17 inches or larger) into a half-moon shape as shown in Figure 3. Tape the foil to the cardboard and bend it into a cone shape. Bend a hanger into the shape shown in Figure 4 and secure the microphone and cone to the hook of the hanger. Position the microphone inside the cone so it faces the point. Tape the cone closed.

Turn up the volume control, and you'll be able to detect distant sounds with the microphone that can be heard through the earphone.

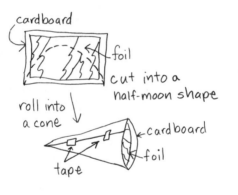

FIGURE 3

FIGURE 4

Give a Shout-out:
Make a Sneaky Megaphone or PA

Are you tired of yelling in your house like a drill sergeant? With this project, you'll learn how to prevent the need to sound off to family members.

Most people believe a cassette tape recorder can only be used to record and play back voices and music. If the motor fails or if it eats tapes, they will usually throw it away. This project will show how it can be used for other sneaky purposes.

You've been in stores where an announcement over a public address system is being played. You can perform the same function with a common tape recorder and a small speaker.

Sneaky fact: A tape recorder sends sound signals to its earphone jack while it is in the RECORD mode. You can prove this by listening with an earphone while you record a tape. If, instead of an earphone, a small speaker is connected to the earphone jack, a PA effect will be heard while recording.

What's Needed
- Tape recorder
- Wire
- Small speaker
- Earphone plug
- Separate microphone (optional)

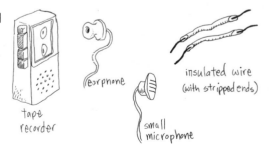

earphone

insulated wire
(with stripped ends)

tape
recorder

small
microphone

What to Do

When placed in the RECORD mode, a tape recorder will amplify
sounds from the microphone and send signals to the tape head.
It also sends audio signals to the earphone jack. To use this function,
place a cassette in the recorder, as shown in Figure 1. Then use
the plug from an old earphone cable and attach its wires to a small
speaker. Any small speaker will work; one from a car sound system
or an old TV, a Walkman speaker, or one from a nonworking radio
or tape player. Attach the wires to the earphone plug.

After connecting the speaker to the earphone jack, place the
cassette in the recorder and press the RECORD button. Then press
the PAUSE button. This will stop the motor from turning and will
save battery power.

Now speak into the microphone, and your voice will be
amplified and sent out through the speaker. Be sure to keep the
speaker pointed away from the recorder to prevent feedback;
see Figure 2. Congratulations! You've just made a portable
public-address system.

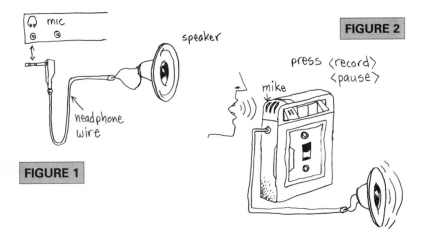

If you add a longer pair of wires, you can speak from one room and be heard in another, much like a one-way intercom (see Figure 3). You can make announcements, play prerecorded tapes, or provide long-distance music and news broadcasts.

Howdy!

FIGURE 3

Secret Agent:
Mr. Wireless
Has Countless Uses

Tired of missing phone calls or a doorbell ring when you leave a room? Fortunately, a sneaky remedy is available.

One of the most versatile toys available is the Mr. Microphone wireless broadcaster. Most people purchase it for their kids so they can sing on an FM radio, but this project has many other applications.

Mr. Wireless, as we'll call it, is essentially a wireless FM radio station with a broadcast range of up to 150 feet. A nearby FM radio can receive sound or music from the microphone. A small screw on the unit allows you to select the broadcast frequency. Here are just a few of its sneaky uses (see Figure 1).

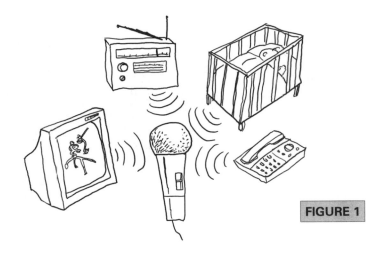

FIGURE 1

- Place a Mr. Wireless in a room near your baby and you can monitor the child with an FM radio.
- Monitor the doorbell or telephone by placing a Mr. Wireless in the vicinity.
- You can also use a Mr. Wireless to monitor a television, shortwave radio, or police scanner.

The Mr. Wireless device can be easily modified to accept a direct audio signal from an earphone jack. This will allow you to use it to play the output of a portable scanner, MP3 player, cassette player, or CD player in an automobile or boat. The car stereo will receive the Mr. Wireless broadcast and send the amplified sounds to the car speakers.

What's Needed
- Mr. Microphone toy
- Earphone plug
- Tape

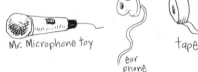

Mr. Microphone toy

ear phone

tape

What to Do
Figure 2 shows a typical Mr. Wireless toy. Remove its sponge cover and remove the case screw(s); some covers will require that you slide and twist the case until it separates into two parts.

Then remove the small microphone from the case.

Cut the wires and strip the ends of insulation (see Figure 3). You can use the plug end from an old earphone, attach it to the microphone wires, and tape them together. Push the plug jack through the top hole, replace the case cover, and screw it together.

Place fresh batteries in the Mr. Wireless, turn it on, and connect the earphone plug to the earphone of the device you

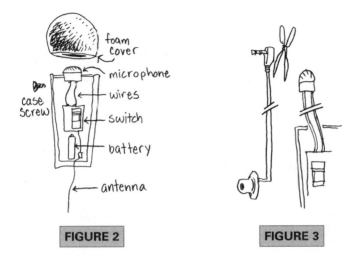

FIGURE 2

FIGURE 3

want to play. Tune the car stereo to the broadcast frequency of the Mr. Wireless and turn on the portable audio player (MP3 player or cassette or CD player). Keep the volume low on the device so it does not produce audio distortion.

Miniaturizing Mr. Wireless:
Use Him in
Remote Places

They say that size matters, but that depends on what your goal is and what you have to work with. If you use the Mr. Wireless device frequently, you'll probably want to miniaturize it for more portable concealed uses. The original package is relatively large, to fit the batteries inside, but it can be made smaller.

With a smaller version, Mr. Wireless can be left at various locations for remote monitoring. Imagine the fun and sneaky applications!

What's Needed
- Mr. Microphone toy
- Lithium or other high-capacity watch battery
- Small plastic candy box or mint container
- Tape

Mr. Microphone toy

tape

plastic candy box

lithium battery

What to Do
Wireless microphone-type toys vary in design, so the following directions are general in nature.

Remove the screw(s) holding the Mr. Microphone case together. In some instances this will require sliding the two case pieces apart; see Figure 1. Then remove the circuit board, ON/OFF

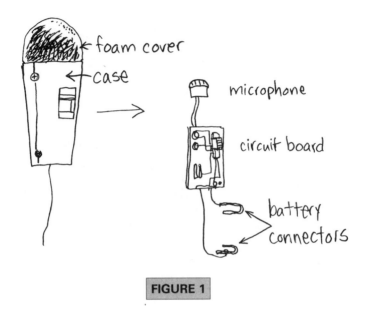

foam cover
case
microphone
circuit board
battery connectors

FIGURE 1

switch, microphone, and battery connectors from the case. Note which connectors are attached to the positive (+) and negative (–) battery terminals.

Assemble your Mr. Wireless parts inside a mint container or other small box, as shown in Figure 2. Tape the lithium watch battery to the connectors and test it with an FM radio to ensure the proper polarity. If it doesn't work, reverse the connectors on the battery. Now your Mr. Wireless is ready for undercover operation.

Keep in mind that miniaturizing the Mr. Wireless device has one disadvantage: The smaller battery used for the modification will have a shorter life.

Remember: The mini Mr. Wireless can be used just about anywhere. But don't forget to bring an FM radio along to monitor its broadcasts.

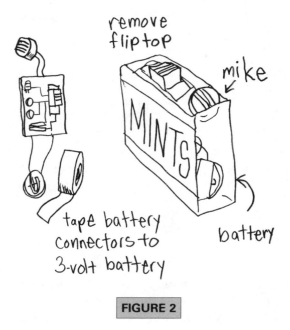

remove fliptop

mike

MINTS

tape battery connectors to 3-volt battery

battery

FIGURE 2

Got a Toy Car?
Make a Power Room Door Opener

Who among us hasn't dreamed of having a power door opener as seen in sci-fi and spy movies? This project will show you how to use a small toy car to do the trick. A small wire-controlled car has enough power to push and pull a typical room door back and forth if you know the super-sneaky way to install it.

What's Needed

- Wire-controlled toy car
- Velcro tape, adhesive-backed
- Screwdriver
- Pliers

Velcro tape

pliers

screwdriver

wire-controlled toy car

What to Do

This project requires a small wire-controlled toy car, *not* a radio-controlled version. This is to prevent the batteries from running down. (With a radio-controlled car, the remote control and the car's internal receiver have to be in the ON mode, and this drains batteries.)

First, remove the body shell from the toy car with a screwdriver. Then remove the front wheel and axle, as shown in Figure 1.

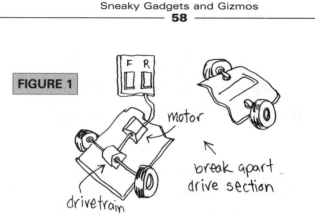

FIGURE 1

motor

break apart
drive section

drivetrain

Now, using the Velcro tape, attach the car near the bottom end of the door (see Figure 2).

Using the remote control, see if it can push the door open or closed. If not, reposition the car for more traction. When the proper position is found, you will be able either to move the door with your hand or let the car do it.

Optionally, you can break off the entire front part of the

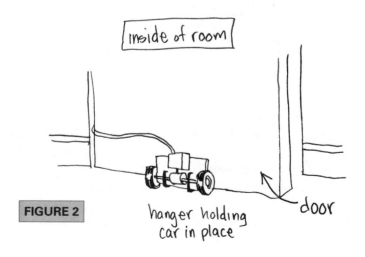

inside of room

FIGURE 2

hanger holding
car in place

door

chassis so that it takes up less space and cover it with materials for a more appealing look. Mount the remote control outside the door as desired (see Figure 3).

velcroed
to wall

F R

FIGURE 3

outside of room

BONUS APPLICATION IDEA

Substitute a magnetically sensitive sneaky switch for the remote control's switch, and your door can be opened with a power ring. (See the "Invite the Power!" section, earlier in Part II, for details.) Now that's futuristic!

Irrational Public Radio:
Put It Together
from Scratch

Ever wonder how a radio works? Would you believe that you can make a sneaky radio with nonelectronic items found in every home? It's true!

Making crystal radios was a popular activity a few generations ago. Even now, many parents show their kids how to make a radio at home that doesn't require AC power or batteries. And there is no danger of getting an electric shock. Although only a few stations will be received, there's nothing like the feeling of tuning in a station for the first time on a radio you put together from scratch.

Normally, building a crystal radio requires electronic components, such as a crystal or germanium diode to act as a detector. This involves going out to purchase a crystal radio kit or hunting for the separate electronic components.

The Irrational Public Radio is made completely out of everyday things. You should be able to receive a strong radio signal in most places in the country. You can use wire from an old telephone cord for the coil, antenna, and ground wires. Instead of a crystal or a diode you will use a penny and a twist-tie!

Radio Fun-damentals

Radio stations mix the audio (sound) signal with a carrier signal. The carrier is a high-frequency radio wave that allows for long-range transmission.

Radio receivers have four basic sections: a receiving section, using an antenna and ground wires to capture radio signals; a tuning section, using a coil of wire, to focus on a specific signal; a detector section, using a diode, to separate the audio signal from the carrier wave; and an audio output section, using a speaker or earphone to convert the audio signal into sound.

Crystal and diode detectors are electronic devices that conduct electricity in one direction better than the other. Crystals were once used in the detector sections of radio receivers, but diodes are used now. Since crystals and diodes are not considered everyday things, our sneaky radio will use a substitute detector made from a penny and a twist-tie.

What's Needed
- Toilet paper tube or small bottle
- Telephone cord, 25 feet or longer
  (or other thin wire)
- Crystal earphone (not a Walkman or
  cell phone headset or in-the-ear earphone)
- Penny
- Paper clips
- Twist-tie
- Cardboard (from a shoe box or food container)
- Pliers with insulated handles
- Screwdriver

toilet
paper

paper twist-
clip or tie

cardboard

pliers

screwdriver

What to Do

Remove the outer insulation from the telephone wire and separate the wires (Figure 1). One of the wires will be used for the antenna. Mount it as high as you can in your room. Lay the other end of the wire on the cardboard with a paper clip, as shown in Figure 2. Strip the insulation from the end of the wire and attach it to the paper clip.

Connect another wire, with its end stripped, to a metal water pipe under a sink in your kitchen. This will be the ground wire. Its other end connects to paper clip 2 as shown in Figure 2.

To build the radio tuning coil, remove two inches of insulation from the end of a third wire and put this stripped end through the paper tube. Wind about eighty turns around the paper tube. Secure the wire with tape if necessary. Insert the end of

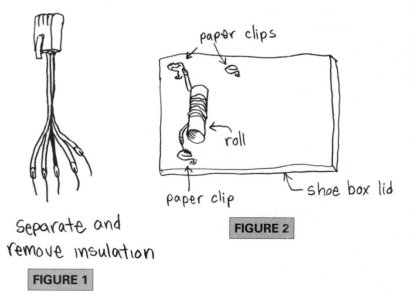

paper clips

roll

shoe box lid

paper clip

FIGURE 2

Separate and remove insulation

FIGURE 1

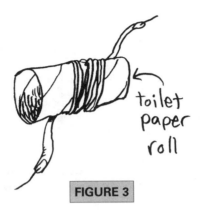

toilet
paper
roll

FIGURE 3

the wire into the other hole on the tube, as shown in Figure 3. Connect both wire ends to paper clips 1 and 2.

Strip an inch of insulation from both wire leads of the earphone. Connect one of the earphone wires to paper clip 1.

Scratch the surface of the penny with the screwdriver and, using pliers, hold it over a flame for three minutes until a black spot appears. After it cools, slip the penny securely under paper clip 2 (see Figure 4).

Strip off the insulation from the twist-tie and cut one end into a sharp point. Bend the twist-tie into a **V** shape and mount the dull end to paper clip 3. The pointed end should press firmly against the burnt surface of the penny, as shown in Figure 4.

Connect the remaining earphone wire to paper clip 3. Your Irrational Public Radio is completed.

Put on the earphone and slowly move the tip of the twist-tie across the surface of the penny until you hear a radio station. Be sure the tip of the twist-tie has enough pressure on the coin to maintain contact. If necessary, bend it into position again (see Figure 5).

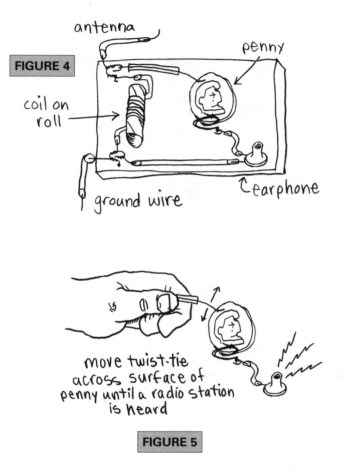

FIGURE 4

antenna

penny

coil on roll

ground wire

earphone

move twist-tie across surface of penny until a radio station is heard

FIGURE 5

BONUS APPLICATION

If you want to amplify the radio's audio signal and allow others to hear it, you can connect it to a tape recorder.

What's Needed
- Tape recorder
- Small speaker
- 2 earphone plug ends (⅛-inch plugs)

tape recorder

small speaker

earphone

What to Do

When placed in the RECORD mode, a tape recorder will amplify sounds from the microphone and send signals to the tape head. It also sends audio signals to the earphone jack.

To use this function, place a cassette in the recorder. Then, using the plug from an old earphone cable, attach its wires to a small speaker. (Any small speaker will work; including one from a car sound system, an old TV, a Walkman, or a nonworking radio or tape player.) Attach the wires to the plug.

Now connect the other earphone plug end to the connections where the radio's earphone would be (paper clips 1 and 3).

After connecting the speaker to the earphone jack, place the cassette in the recorder and press the RECORD button. Then press the PAUSE button. This will stop the motor from turning and will save battery power. You will hear the radio signal emanate from the small speaker, and everyone in the room can enjoy the sounds from the "penny radio."

Tips

A crystal radio will usually operate best at night. Under some combinations of conditions you may receive no stations at all. Be patient and try again.

If no sounds are detected, not even a slight hum, check all wire connections to the paper clips and reposition the wire attached to the water pipe.

You can launch an antenna wire high up into a tree with a rock or toy dart gun. Wrap the end of a 50-foot wire to the end of a dart and shoot it toward high branches.

You can send the radio's output into the microphone input of a tape recorder and listen to it with a speaker connected to the earphone jack.

Other detector materials that can be used include a silver gum wrapper, a spring from a ballpoint pen, a rusty nail or washer, a rusty set of pliers. Remember: the detector's two parts must be two *unlike* metals.

More information on crystal radio building can be obtained from the following Web sites:

boydhouse.com/crystalradio
howstuffworks.com
midnightscience.com
scitoys.com

Con Air:
Convert Your Radio into an Aircraft Broadcast Receiver

"**F**light Four-eight-three to LAX tower . . . in need of immediate assistance. We're experiencing an emergency situation."

The VHF aircraft band is filled with fascinating and sometimes critical communications from commercial airliners, the military, and private aircraft. Many aircraft-scanner listeners enjoy the fun and excitement of tuning in tower-to-air conversations that are missed by the general public. Although aircraft radios command premium prices, there is a way to enjoy aircraft broadcasts for free. In fact, if you have an AM/FM radio, you already have an aircraft radio!

Believe it or not, you can easily convert virtually any AM/FM radio into an effective aircraft band receiver. The AM band will remain unchanged. The FM band will have limited reception after the conversion, but it can quickly be returned to its original condition with a small standard screwdriver.

This project should use a small nondigital battery-powered radio. Not too small—the ultra-tiny Walkman radios can be difficult to open and adjust. But first, some background information on how radios work and what the conversion will do.

Broadcast radio stations transmit a combination of two signals: the audio signal, which includes voice or music, and the carrier signal, which is designed to travel long distances and "carry" the audio signal with it. Voices or music affect (or modulate) the

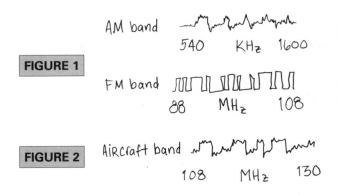

AM band

540   KHz   1600

FIGURE 1

FM band

88   MHz   108

FIGURE 2   Aircraft band

108   MHz   130

audio signal, either by varying its strength, known as *amplitude,* or its *frequency.*

Using voice and music to vary the *amplitude* of the carrier signal is called amplitude modulation—AM. Varying the *frequency* of a carrier signal is the defining feature of *frequency* modulation (FM). See Figure 1 for an illustration of these principles.

An AM radio is designed to tune to a specific range of broadcast stations, discard the carrier signal, and leave the audio signal. Then, just the strength (amplitude) of the signal is forwarded to the radio's speaker or earphone. On the other hand, in an FM radio just the varying *frequency* of the signal is sent to the radio's speaker or earphone.

Fun Fact 1: AM radios are more sensitive to static because radio interference is a result of the strength of radio waves emanating from lightning or home appliances. FM radios are designed to filter out changes in the strength of the signal, and that's why they produce superior audio quality.

Aircraft radio signals are broadcast on the AM band at a range just above the standard FM band; see Figure 2. What you'll do is convert the FM section of the radio to AM and also allow it to receive signals within the aircraft band. The best thing is that the conversion is simple and does not require any parts!

Fun Fact 2: Television signals are both AM and FM; the picture information is broadcast in AM, and the sound signal is broadcast in FM. The two signals are combined and "ride" on the carrier signal. The next time you experience electrical inter-ference in your house, you'll notice that the sound of the TV set is not affected as much as the picture.

What's Needed
- Portable AM/FM radio
- Small flat-blade screwdriver

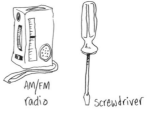

AM/FM radio   Screwdriver

What to Do
Test the AM/FM radio to ensure that the battery is strong and that local radio stations can be received loud and clear. Open the back of the radio so you can see the parts on the main circuit board.

Note: Some radio cases snap or slide open while others require you to remove a screw.

Examine the items on the circuit board and look for the part under the tuning dial that selects the stations. This is the variable capacitor. You'll also see two or three small square metal compo-nents. They are filter transformers. Their purpose is to filter out noise and static on the FM band. In essence, they filter out AM signals. Look even closer at the top of the filter transformers, and you'll see a screw slot to allow adjustment. Also, one of the filter transformers will have two or three small glass diodes next to it. A diode looks like a clear slender bead with wires on each end. This is the filter transformer that we will adjust.

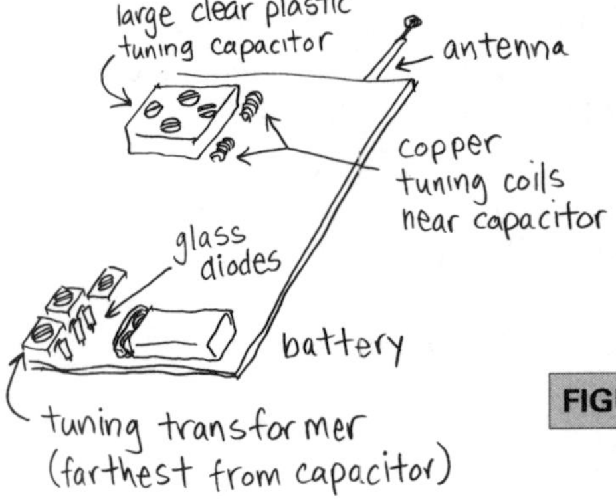

large clear plastic tuning capacitor

antenna

copper tuning coils near capacitor

glass diodes

battery

tuning transformer (farthest from capacitor)

FIGURE 3

Note: In some radios you may not see the diodes near a tuning transformer, but the adjustment can still be performed. See Figure 3 for an illustration of a typical radio and its component parts.

The first step is to turn the radio on and switch to the FM band. Tune to a spot between stations until you hear a background hiss. Now place your screwdriver in the top of the filter transformer and turn it until the hiss gets as loud as possible. When the hiss is at its highest volume, you have just converted the FM portion of the radio so it can receive AM signals.

One more step is required—you must extend the broadcast range of the FM band. To do this, look at the radio's tuning dial and notice the large square tuning capacitor on the main board. There will be two small copper wire coils next to it. With the screwdriver, spread the small coil windings apart as much as possible without letting them touch another part on the board (see Figure 4). By spreading the coils apart, you have extended the broadcast range from 88 megahertz to 108 MHz and above.

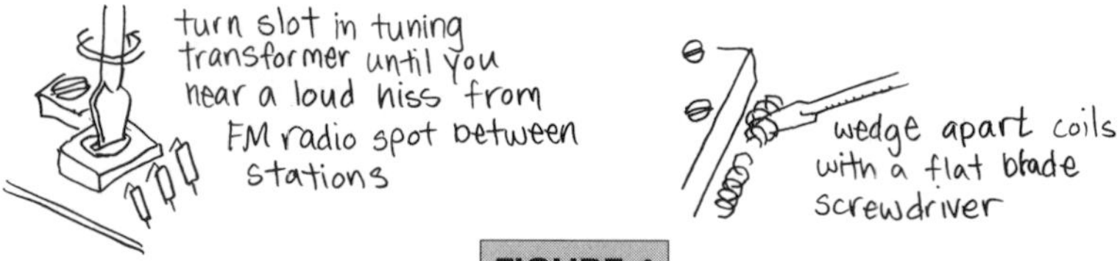

turn slot in tuning transformer until you near a loud hiss from FM radio spot between stations

wedge apart coils with a flat blade screwdriver

FIGURE 4

Congratulations! Now your radio can receive signals within the aircraft radio band.

Next, tune the dial up and down the FM band. You should notice that the station locations have changed, now that the radio is set to receive broadcasts on the aircraft radio band. Take the radio near a local airport and tune across the band. Aircraft messages—from automatic runway signals to tower communiqués—should be audible while tuning the dial.

You can easily convert the radio back to its original state by reversing the two-step process. First, push the small wire coils that are next to the tuning capacitor back to their original positions. Second, dial the radio to a section between stations and retune the filter transformer until the hiss sound is at a minimum.

The next time you visit an airport, take your multiband radio with you and enjoy the fun and intrigue of eavesdropping on aircraft band conversations.

Part III

Security Gadgets and Gizmos

Most people like to believe that break-ins and other misfortunes won't happen to them. But ignorance is not always bliss! Nowadays, home, apartment, and hotel security matters are a fundamental concern. This section provides protection devices that can be rigged up to foil assaults on person or property using paper clips, rubber bands, and other household odds and ends.

First, you'll become skilled at using the most innocent-looking items to secure your personal things, like a letter, wallet, or purse. Next, you'll learn how to obtain "see-behind vision" and how to make devices to detect if your doors or windows have been tampered with. Also included are sneaky projects that show how to use a tape recorder to thwart intruders and how to rig a disposable camera to identify burglars by taking a thug shot.

Sneaky Ways to Thwart Break-ins:
Protect Your Fortress from a Man of Steal

Sure, traveling has its charms, but with exotic environments sometimes come mysterious and unexpected dangers. It doesn't happen often, but when you stay overnight away from home you probably worry about someone entering your room without your knowledge. The following projects provide a trio of portable and quick-to-set-up security gimmicks for use at home, on a trip, or for an unforeseen stay in a foreign location to thwart or detect window and door break-ins at a moment's notice.

What's Needed
- Rubber bands
- Bubble-wrap material
- String or wire
- Cardboard
- Small bell or chime

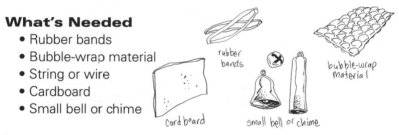

rubber bands

bubble-wrap material

cardboard

small bell or chime

What to Do
To be warned when someone enters a room, place bubble-wrap material under a mat or towel near the entrance. Or place a small bell or chime on the door or window (see Figure 1).

Place an External Sneak Detector (see "Thwart Thieves" later in Part III for details) against a window handle or on a doorsill so it will produce a loud popping sound when activated (see Figure 2).

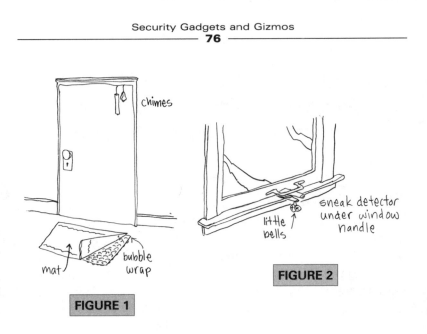

chimes

mat

bubble wrap

FIGURE 1

little bells

sneak detector under window handle

FIGURE 2

Optionally, use whatever unbreakable but noisy items are available to place against a door or window to alert you when they are displaced.

HOW TO PREVENT BREAKING AND ENTERING

What's Needed
- Wire or strong nylon thread
- Broom, mop handle, plunger dowel, or chair
- Towels
- Tape or rubber bands
- Duct tape

broom or mop handle

rubber bands

tape

wire or nylon thread

What to Do

To protect a door from opening, place a long object, like a broom, mop, or chair, under the doorknob so it's wedged in tight. If necessary, use towels or small pillows to anchor the object to the floor to prevent slippage (see Figure 1).

Even strong nylon thread can prevent a door from opening if it's wrapped tightly from the doorknob to an adjacent window handle or wall light-plate fixture screw. Apply duct tape to all corners of the door to further prevent a break-in (see Figure 2).

Similarly, a window can be secured with an object to prevent it from sliding in its track, as shown in Figure 3.

If a single long object is not available, use shorter ones, like aerosol cans or slender bottles. Position all of the objects end to end and secure them with tape or rubber bands to keep them in line (see Figure 4).

Although a determined burglar can breech these entry inhibitors, they give you valuable time to react and call for help. A chime can be taped to the windows and doors to alert you that a break-in is in progress.

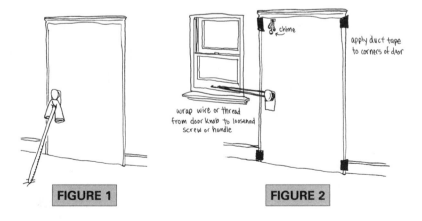

chime

apply duct tape
to corners of door

wrap wire or thread
from door knob to loosened
screw or handle

FIGURE 1   FIGURE 2

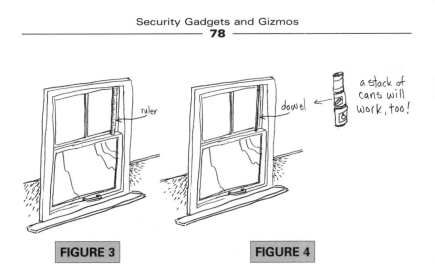

FIGURE 3 **FIGURE 4**

PHANTOM MENACE NOISEMAKERS

Your home audio/video system is probably your pride and joy.
Others might enjoy it too. But they may not spend the time and
money shopping at the places you did. They may take the direct-
theft route to aural delight.

What's Needed

- Wire
- Aluminum foil
- Cardboard
- Radio, tape recorder, or
  battery-powered clock radio
- Ballpoint pen spring

What to Do

You can cause a door or window opening to activate a battery-
powered radio, a tape player, or a battery-powered alarm clock

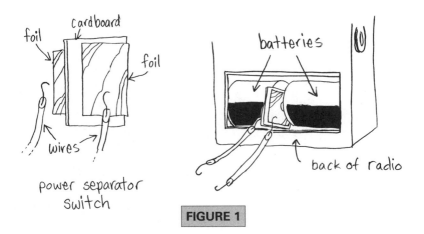

foil
cardboard
foil
wires
power separator switch

batteries
back of radio

FIGURE 1

using a power separator switch and contact sensor. Any battery-powered toy or gadget that makes noise will work. The power separator switch, made from foil, wire, and cardboard as shown in Figure 1, is placed between batteries in a radio, tape player, or noise-making toy.

First, tape a piece of aluminum foil at the side of the door near the hinge. Then remove the spring from a ballpoint pen. Tape the spring on adjacent area, near (but not touching) the aluminum foil strip on the door. Bend and mount the spring so that the door must be open about one third of the way before it touches the foil on the door.

Next, attach wires to the spring and the foil and connect them to the wires on the power separator switch (see Figure 2). Now, when the door is opened, the spring contacts the foil, which completes the circuit, allowing the radio or tape player to obtain power from the batteries. Be sure to set the volume high enough to alert you that a door or window has been breeched.

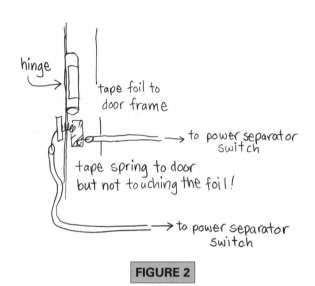

hinge

tape foil to door frame

→ to power separator switch

tape spring to door but not touching the foil!

→ to power separator switch

FIGURE 2

BONUS APPLICATION:
PHANTOM MENACE NOISEMAKER

Sure, you could install a costly and elaborate home security device that blares an ominous warning over a PA system. But you can make one with items found in virtually every household.

Using the same parts shown in the "H_2O No!" project described a little later here in Part III, you can make a cassette tape recorder protect your house or apartment in a sneaky way. By prerecording a stern, ominous message on the tape and playing it back through a large speaker, your living quarters can appear protected by an expensive alarm system and give the impression that a live security team is on the way.

Substitute a playing card (or other laminated piece of card-board) with a thin wire or nylon thread attached (see Figure 1).

Attach the other end of the wire to a window handle or door or to a TV, stereo, or other appliance (see Figure 2). When it's moved, the card will separate from the paper clips and turn on the recorder and it will play the tape.

You can have the sound emanate from the recorder's speaker or, for a greater effect, you can plug in a remote speaker (obtained from an old TV or car stereo or other entertainment device) and connect it with an earphone cable and wire; see Figure 3.

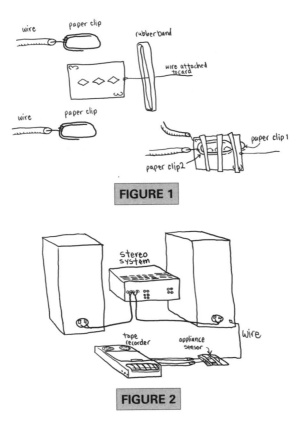

FIGURE 1

FIGURE 2

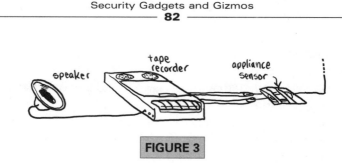

FIGURE 3

Here's a sample recorded script: *Stop. You have entered unlawfully. Security personnel have been notified and are on the way. Leave immediately.*

BONUS APPLICATION 2:
X COMMUNICATE

The sneaky alarm just described can activate an X-10® telephone dialer (or one from security system dealers). When an alarm is activated, you can be notified by pager, phone, or e-mail!

For more information about security dialers, see Resources at the back of the book.

Foam Alone:
Make a Sneaky Fire Extinguisher

With a little planning and two items found in your kitchen, you can have a lifesaving tool available for emergencies. Having a fire extinguisher on hand gives you peace of mind and can prevent costly damage to your belongings (and possibly save a life). If you ever need a compact and portable fire extinguisher, you can make a sneaky version in just a few minutes using household materials.

What's Needed

- Baking soda
- Vinegar (or lemon juice)
- Jar or plastic soda bottle, 1 pint or larger
- Tissue
- Tape or a rubber band

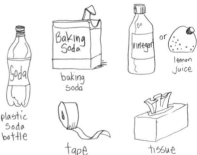

What to Do

Fill the bottle halfway with vinegar (or lemon juice). Then form the tissue into a cup shape and poke it into the mouth of the bottle, as shown in Figure 1, holding the sides of the tissue over the rim of the bottle.

Still holding the sides, pour baking soda into the tissue and then place a rubber band or tape around the bottle opening to prevent the tissue from dropping into the bottle; see Figure 2.

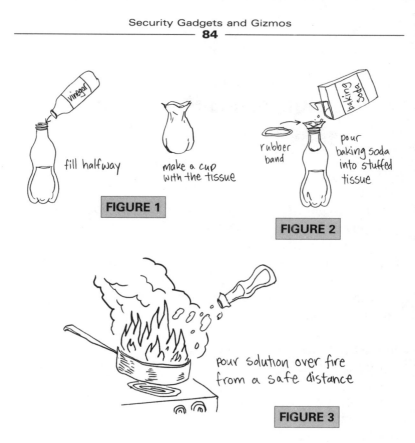

fill halfway

make a cup with the tissue

rubber band

pour baking soda into stuffed tissue

FIGURE 1

FIGURE 2

pour solution over fire from a safe distance

FIGURE 3

Replace the cap on the bottle or place another tissue and rubber band or tape on top of the bottle while it's in storage.

When needed, shake the bottle vigorously and remove the cap. A fire-retardant foam will bubble up, to be cautiously poured over a fire. See Figure 3.

Caution: Be particularly careful when in the presence of fire, especially an oil or grease fire.

Gain Sneaky
See-behind Vision

Forget X-ray vision, heat vision, and microscopic vision. In the real world, whether you're at an ATM or opening your car door, what counts is "see-behind" vision.

Many assaults take place when a person is distracted, and it's impossible to keep looking around while you're fiddling with keys and cards. Luckily, a sneaky remedy is at hand.

What's Needed
- Small mirror or reflective material, about 1½ inches square
- Duct tape
- Large paper clip

tape

mirror

paper clip

What to Do
Bend the large paper clip into the shape shown in Figure 1 and tape it tightly to the small mirror. If the edges of the mirror are

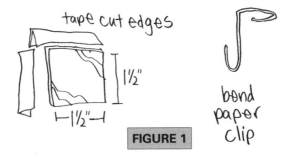

tape cut edges

1½"

├─1½"─┤

FIGURE 1

bend paper clip

sharp, tape them carefully also. Then attach the mirror to a cap or glasses with the other end of the clip, as shown in Figure 2. Note: Adjust the position of the mirror, by bending the paper clip, to improve clarity if required.

You can use this sneaky vision device three ways: it can be carried in your hand; it can be attached to a cap; or it can be clipped to eyeglasses. Now go ahead and use your See-behind vision for safety and fun.

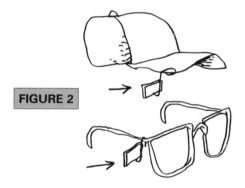

FIGURE 2

Industrious Light Magic:
Make a Sneaky Light in a Pinch

There's an irony in being caught without a light source in a car, when there are nearly twenty light bulbs in the average vehicle plus a 12-volt battery under the hood. The only thing required is some wire to connect the battery to the bulb.

Many automotive emergencies happen at night, and most people drive without a working flashlight. You could be caught without ample light to see where a problem is. Light may be needed to find a part on the ground. If you should need to leave the vehicle on a road, a light can be used to signal for help or reveal yourself to motorists so you won't be hit.

This project illustrates two methods of making a small working light from everyday things.

MAKE AN UNDER-HOOD LIGHT

What's Needed
- Small bulb
- Wire

flash
light
bulb

insulated wire
(with stripped ends)

What to Do
Wire can be obtained from speaker wire connectors or from some other noncritical source in the vehicle. (See "Getting Wired" in Part I for more sources of emergency wire.) As shown in Figure 1, wrap the bare wire around the bulb's side and hold the other end on the battery's positive (+) terminal. Then touch the bulb's bottom to a metal part away from the battery under the hood.

FIGURE 1

If you have two long lengths of wire available, you can attach one end to the bottom of the bulb and the other to the side and have a more flexible light.

MAKE A PORTABLE LIGHT

There's also a way to make a sneaky light for use *away* from the vehicle. When using an automotive bulb, you can use a small 12-volt battery that some portable alarm remote controls use. Otherwise, if you have a flashlight bulb (and the flashlight batteries are depleted) you can use its bulb with batteries you may have on your person.

Note: Do not exceed the bulb's recommended voltage requirement. A, C, D, and AA batteries all supply 1.5 volts. So if a flashlight uses two D-cell batteries, the bulb requires 3 volts to operate. You can then use two C or AA or small watch batteries to light the bulb. Batteries can be found in a pager, a car alarm's remote control, a toy, a garage-door opener, or a watch. (In a remote survival situation, batteries from fruits can be used; see "More Power to You" in Part I for details.)

What's Needed

- Battery (that meets bulb voltage requirement)
- Bulb
- Wire

battery

flash light bulb

insulated wire (with stripped ends)

What to Do

You will need a small length of wire. If no standard wire is available, alternative everyday things that can be used include a paper clip, a twist-tie, a metal chain or earring, a ballpoint pen spring, keys, speaker wire, or aluminum foil from a snack bag or coffee-creamer-container lid. See Figure 2.

If more than one battery is required, you can wrap paper around them and hold the paper in place with tape, wire, a rubber band, string, or a twist-tie.

With the bulb, wire, and batteries available, attach the parts together as shown in Figure 3. If you use small batteries, try to limit your use of the sneaky light to save power.

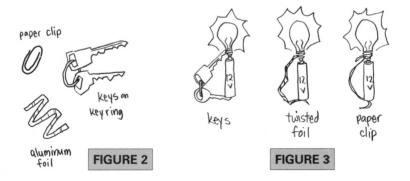

paper clip

keys on key ring

aluminum foil

FIGURE 2

keys

twisted foil

paper clip

FIGURE 3

H₂O No!:
Make a Sneaky Flood Alarm

Few events are more distressing than returning home to a flood situation. This project shows how to construct an easy-to-make sneaky flood alarm for your home from everyday things. You'll need only an inexpensive battery-powered clock or AM radio and a few other items to complete the alarm.

What's Needed

- Any fast-dissolving tablet
- Rubber bands
- Wire
- Aluminum foil
- Cardboard
- Paper clips
- Battery-powered AM radio or alarm clock

What to Do

You can cause a seepage of water on the floor to activate a battery-powered radio, using a power separator switch and water sensor. The power separator switch, made from foil, wire, and cardboard, is placed between batteries in a radio, as shown in Figure 1. The cardboard separates the aluminum foil contacts from completing the circuit.

The foil contacts are attached to two paper clips, which are separated by a fast-dissolving substance—seltzer tablet, sugar cube, aspirin—in the sneaky flood switch. The rubber band applies

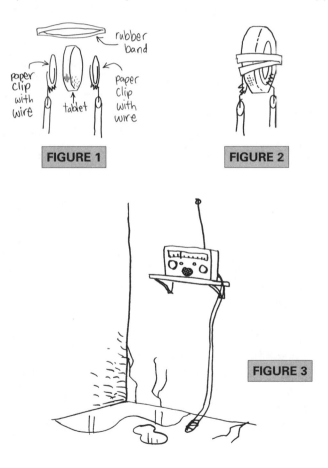

FIGURE 1

FIGURE 2

FIGURE 3

pressure (see Figure 2). When water surrounds the tablet, it will dissolve, causing the paper clips to make contact (because of the pressure from the rubber bands). This allows the radio to obtain power from the batteries. The radio, which is tuned to a strong station and set to maximum volume, will turn on with a loud roar, alerting you to the danger (see Figure 3). If you use a battery-powered alarm clock, set the alarm to activate as soon as it's turned on.

Test how fast the item you use as the water sensor will dissolve. For example, a few oral breath strips placed between the paper clips will dissolve in just a few seconds.

If necessary, use longer wires from the sneaky flood sensor so the radio can be placed where you can hear it from anywhere in the house. Another option is to connect more sneaky sensors in parallel to the first one so other areas can be protected as well.

BONUS APPLICATION:
A SNEAKY FIRE ALARM

A flood alarm isn't the only sneaky device that can be made with the power separator and paper-clip sensor. Using the same items shown in Figure 1, you can make a sneaky fire alarm too.

Construct the parts as before but, instead of a water-soluble tablet, place beeswax (or another fast-melting substance) between the paper clips (see Figure 2). Set the fire sensor as high as possible, but not near the corner of a room. Virtually any soft fast-melting substance will work (see Figure 3).

It is a good idea to place a container below the sensor so the melting substance will not collect on the floor.

Note: This device is not intended to replace a standard fire alarm. It can be used as a demonstrational device or until a standard UL-approved alarm is available.

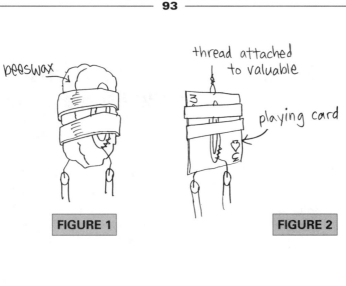

beeswax

thread attached to valuable

playing card

FIGURE 1

FIGURE 2

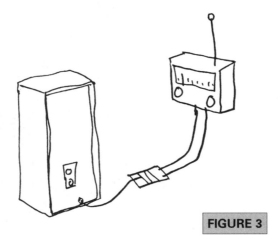

FIGURE 3

Sticky Fingers?
Keep Watch with an
Internal Sneak Detector

Are you tired of people prying in your personal items? If you believe that someone is opening your mail or tampering with your belongings, you can plant an Internal Sneak Detector for verification. If you return and the device has snapped apart, you will know someone has opened the item.

What's Needed

- Standard-size envelope
- Paper clip
- Small rubber band
- Piece of strong cardboard
- Scissors
- Pen
- Pliers

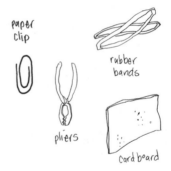

What to Do

You can use strong cardboard from a tissue box, shoe box, or videocassette box. First, cut four small pieces, 3 inches long by 1 inch wide, as shown in Figure 1.

Next, with a pen, punch small holes at the ends of the four cardboard pieces. Then straighten out the paper clip and cut it into four small pieces.

Position the four cardboard pieces and push a piece of paper clip into each hole, as in Figure 2.

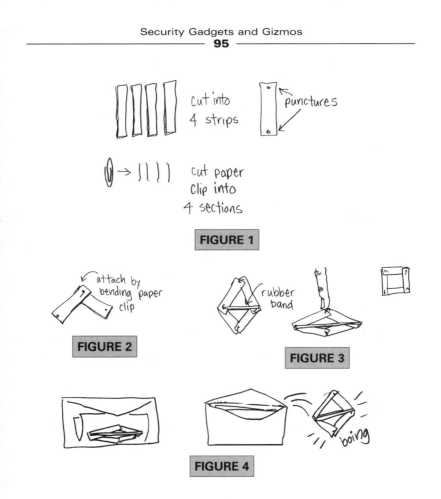

Cut into 4 strips

punctures

cut paper clip into 4 sections

FIGURE 1

attach by bending paper clip

FIGURE 2

rubber band

FIGURE 3

FIGURE 4

boing

Attach the rubber band across two of the ends (see Figure 3).

To use the Internal Sneak Detector, push the other two ends of the rubber band together and place the device into a standard envelope. The rubber band creates spring tension. If the envelope is opened, the detector will spring out, as shown in Figure 4.

You might want to leave a small note, stating that you're aware of the tampering that's been occurring.

Thwart Thieves with the External Sneak Detector

If you use the Internal Sneak Detector, you'll undoubtedly want to try an external version for such larger possessions as a purse, briefcase, book, or laptop computer.

The External Sneak Detector can be placed under an item. It will flip and make a loud snapping sound if the item is moved. You can place a note under it to ward off future mischief.

What's Needed

- Small rubber band
- Piece of strong cardboard
- Scissors
- Glue

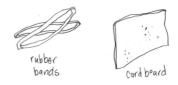

rubber bands

cardboard

What to Do

You can use strong cardboard from a tissue box, shoe box, or videocassette box. First, from four pieces of cardboard, cut two identical replicas each of shapes A and B, as shown in Figure 1. Glue each pair together for added strength.

Shape B is rectangular. Shape A is the same as B except for a thin notch cut into it from one end to its center; see Figure 1. The exact dimensions of A and B are not critical, but Shape A should extend to half the length of the sneak detector and be wide enough for the small rubber band to fit through easily.

Next, glue pieces A and B together along half of their length, as shown in Figure 2. Glue only the unnotched part of Shape A

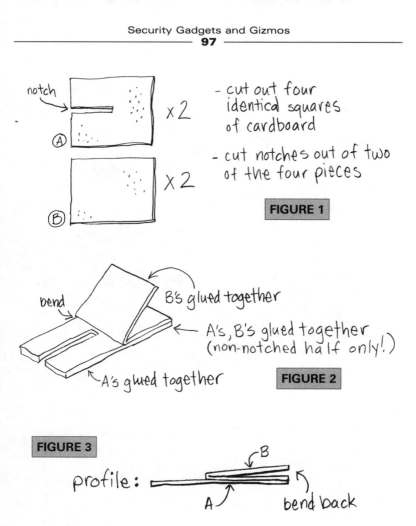

notch

×2

- cut out four
  identical squares
  of cardboard

- cut notches out of two
  of the four pieces

Ⓐ

Ⓑ

×2

FIGURE 1

bend

B's glued together

A's, B's glued together
(non-notched half only!)

A's glued together

FIGURE 2

FIGURE 3

profile:

B

A

bend back

to Shape B. Once the pieces are glued together, bend the
notched half of Shape A back and forth so it flexes (see Figure 3).

Now, with pieces A and B flattened out, attach the small
rubber band. Be sure to guide the rubber band through the notch.
Once it's in place, pull back on the notched half of Shape A until

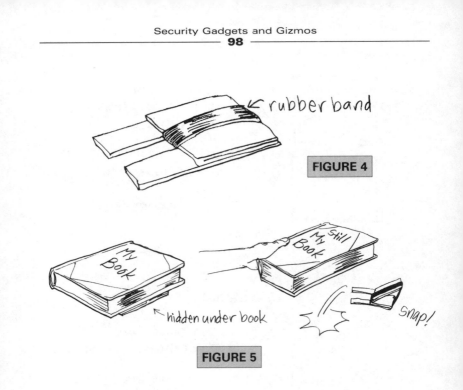

FIGURE 4

FIGURE 5

it bends all the way back. This will stretch the rubber band. If you let go, it will snap back fast and loud like a mousetrap (see Figure 4). If this does not happen, use a smaller rubber band or wrap the band around and through the notch twice.

As seen in Figure 5, the External Sneak Detector, placed under an object, will fly up and snap with a loud pop when the object is moved.

Thug Shot:
Capture Break-ins on Film

A picture can be more valuable than a thousand words . . . in court. If a break-in cannot be thwarted, the next best thing is to record it on film. This project illustrates how to set up a disposable camera so that, if a door or window is breached, you will have a "thug shot" for evidence.

What's Needed

- Disposable camera (non-flash)
- Toy dart gun, with suction-cup tips
- Strong medium-sized rubber band
- Hanger
- Thin strong thread
- Pliers
- Tape
- Cardboard
- Stick-on eyelets

disposable camera

rubber bands

suction cup tip

plastic toy dart gun

hanger

strong thread

tape

pliers

Cardboard

stick-on eyelets

What to Do

This project works when a toy gun, triggered by a door or window opening, shoots its dart into a camera's shutter button.

   Using pliers, bend the hanger into the shape shown in Figure 1 so that it secures the toy gun and the camera. You should be able to slide the gun in and out of the mount so that a dart, with the suction cup temporarily removed, can be easily reloaded.

Place the toy gun so the dart is aimed a few inches from the camera's shutter button (see Figure 2). Attach one end of the string to the toy gun's trigger and the other end to the door (or window) with stick-on eyelets or suction-cup hooks, as shown in Figure 3.

When you set up the thread and paper-clip triggering system, be sure to allow enough room for *you* to slip in and out of the room without setting off the camera. After the correct length and tension of the thread is worked out, when the door (or window) is opened, the thread will pull the toy gun's trigger, causing the dart to shoot the shutter button so the camera takes a photograph.

Since the camera has no flash—a flash would alert the burglar—the room must have sufficient ambient light so the image will be visible on film. Also, since this is a disposable camera, remember to advance the film manually so that it's ready to take a photograph when you leave the room.

Note: This project can also be used in combination with the taping of Phantom Menace Noisemakers described earlier in Part III under "Sneaky Ways to Thwart Break-ins."

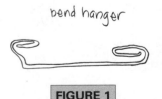

bend hanger

FIGURE 1

↙ suction cup removed

FIGURE 2

eyelets

thread

suction cup

burglar

FIGURE 3

Hide and Sneak:
Secure Valuables in Everyday Things

You've seen movies where a character hides something at home and you think, That's the first place I'd look! Well, this project will illustrate how to choose sneaky locations that are the *last* places a Man of Steal would look. You don't always have access to a safe deposit box or install alarms on all of your possessions. But you can find sneaky hide-in-plain-sight places to frustrate and waste a thief's time.

Selecting this hiding place generally depends on two factors: the size of the item and the frequency of access required. From a package of soap to a tennis ball, a typical home offers a variety of clever hiding places, as shown in Figure 1. Wrapping your valuables in black plastic bags will further prevent discovery.

With enough time, a tenacious thief can eventually find virtually anything you hide. That's why you should have a room entry alarm installed in combination with sneaky hiding places to reduce the time a thief will spend searching for your valuables.

The Resources section at the back of the book provides a list of security companies marketing clever safes that appear to be soda cans, aerosol sprays, and other everyday things.

Speaker safe

pen safe
for emergency
cash

not good for
floppy disks or
items affected by
magnets

inside
handset

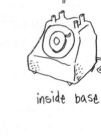

battery compartment
of radio

flowerpot
safe

inside base

slit along
seam

tennis
Dall
Safe

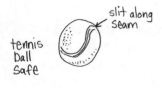

Store
inside
Vacuum
Cleaner
bag

FIGURE 1

Power of the Press:
Use Ordinary Objects as Sneaky Weapons

According to the Bureau of Justice, a violent crime occurs in the United States every five seconds. Being prepared can save your life. You don't have to buy a gun or a can of Mace to protect yourself from attackers. In some instances, ordinary objects can be used effectively as weapons.

First, an important warning: Don't contend with adversaries. You won't lose a fight you don't participate in. If possible, run away as soon as possible and yell for help.

In situations where there is no other way out, you will find that common objects, from a public phone to a magazine, can distract an assailant, keep distance between you, and, if you must strike, reduce injury and protect your flesh and bones. Here's how to defend yourself:

Public phone. If an assailant decides to attack you while you're on the phone, use the handset to hit your attacker on the nose or temple. In a phone booth, use the door to wedge the attacker's arm or leg.

Coins. You can throw a handful of coins at an attacker's face to stun and throw him or her off balance. See Figure 1.

Pens and pencils. Pen and pencils can be used to jab at an assailant's exposed skin and cause enough of a distraction to enable you to escape.

Magazines. By rolling up a magazine and holding it tight, you can strike the face, temple, or ear to temporarily disable an attacker. See Figure 2.

FIGURE 1

WHACK!!

magazine

rolled up

FIGURE 2

One other technique to remember: If the attack has not occurred but is imminent, hold the ordinary object down in a nonthreatening way. The attacker may not realize that you will use it as a weapon.

If you sense that a threatening situation is escalating, look around, grab whatever you can—your shoes, a book, a cup, a package, a box, a small printer: anything you can throw and use it as if your life depends on it.

Part IV

Sneaky Survival Techniques

Who hasn't seen *Gilligan's Island* or the movie *Cast Away* and thought, What would I do if I was lost or stranded somewhere? Can you imagine being marooned without fire, water, tools, weapons, or a compass? What would you do? How would you survive?

In this section you will find ways to stay afloat in sink-or-swim situations. You will learn how to make a fire and collect water, and how to build a makeshift telescope or magnifying glass, use code-signaling techniques, and make a sneaky emergency light.

You'll see how to survive in the cold and hike in deep snow, and you'll learn direction-finding techniques and how to make a compass.

Part IV concludes with crafty ways to devise makeshift weapons and tools.

Even if you feel that it'll never happen to you, review the text and illustrations. Someday your life may depend on it.

Sneaky Emergency Flotation Devices

If you find yourself in a sink-or-swim scenario, what will you do if a flotation device isn't available? Make a sneaky one from everyday things.

When floating in water, the more you try to keep your head above the surface of the water, the more likely you are to sink. Just lie back and keep your mouth above water.

When you attempt to raise parts of your body above the surface, you lose buoyancy. Luckily, however, you can add to your buoyancy with virtually any empty container that holds air. In some instances two or more may need to be secured together.

What's Needed

- Plastic bags
- Gas cans
- Large soda bottles
- Other items that will hold air
- String, wire, a belt, or cloth

plastic bags

gas can

large soda bottles

string

What to Do

To make a flotation device of plastic bags, blow the smallest one up, tie a tight knot, and place it in a larger bag (or bags, if available) as shown in Figure 1 to compensate for small holes. Use these inflated bags as water wings to help stay afloat. Rest on your back with your head up (figures 2 and 3).

Figure 4 illustrates how to connect two water-holding bottles or jars together. You can also use a log if one is available. Be sure

place inflated bags inside as many bags as possible (the more, the better)

FIGURE 1

plastic bags

FIGURE 2

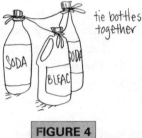

soda bottles

FIGURE 3

tie bottles together

SODA   BLEAC   SODA

FIGURE 4

tie handles together

FIGURE 5

tie legs tight

air will be trapped up in pants

FIGURE 6

the log will float before laying your body on it (not all wood will float).

There are many other flotation devices that you can devise by using some imagination. Just make sure to test their flotation capabilities before trying to use them. Figure 5 shows how to tie bags together.

When no other items are available, your clothing can hold pockets of air to increase your buoyancy. If necessary, remove your pants or shirt, tie the ends of the pant legs or sleeves in knots and scoop air into them. Hold the other end together firmly with your hands and you should be able to ride above the water with little effort (see Figure 6).

Science Friction:
Six Fire-making Methods

Experienced campers know how to start a fire without a lighter or matches, but do you? When lost in the wilderness, being able to make a fire can be a lifesaver, both to signal your location and to use for warmth and cooking.

Everyone has heard that it's possible to make fire by rubbing two sticks together. But exactly how do you do this? What if you don't have two dry and properly shaped pieces of wood? Then what do you do?

This project will illustrate six different ways to start a small fire in an emergency. Some of the methods will work and some will not, depending on the resources available, your skill, and your luck. Review all six methods just in case you need to use them one day.

Before you attempt to start a fire, you must have tinder and kindling materials available and understand how to use them. Many people fail to start fires even when they have good matches!

A fire is built in stages. You need first to cause a small fire spark, with one of the methods shown below, to ignite your tinder: small dry items like tissue paper, dead grass, twigs, leaves, lint, or currency. Blow on the tinder carefully, so that it stays lit and grows into a larger fire. Then add kindling—sticks, branches, or thick paper—(*very* carefully, so that you do not suffocate the flame) to keep the fire going. When the kindling is burning, you can add larger logs or other fuel.

METHOD 1:
MAKE A FIRE PLOW

What's Needed

- Hard stick with a blunted tip
- Flat piece of wood
- Tinder
- Kindling
- Knife or sharp-edged rock

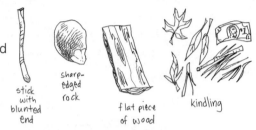

stick with blunted end

sharp-edged rock

flat piece of wood

kindling

What to Do

Using a knife or a sharp rock, scratch a straight indentation in the center of the flat piece of wood about the same width as

FIGURE 1

the blunt stick. Arrange the tinder so air can easily circulate and set it at the foot of the piece of wood, as shown in Figure 1.

Then, in a kneeling position, hold the flat piece of wood between your knees at an angle and move the stick rapidly back and forth in the indentation until friction ignites the fibers of tinder at the base. Mix in more tinder material and fan the smoke until a small fire starts. To keep the fire going, carefully add kindling material.

METHOD 2:
SPARK GENERATION

What's Needed
- Knife or steel
- Sharp-edged rocks
- Tinder
- Kindling

sharp-edged rock

kindling

What to Do
Use this method with very dry tinder material in a secluded nonwindy environment. Depending on what items are available, strike two rocks together to create a small spark close to tinder material (Figure 2). If a spark catches the tinder, you will see a glow. Carefully blow it so it turns into a small flame. Fan the material until it starts to smoke and burn (as in Method 1).

If you have an item made of steel, like a knife, scrape it against various rocks until a spark appears.

FIGURE 2

METHOD 3:
MAKE FIRE WITH A BATTERY

You can use the battery from a car or recreational vehicle (or batteries from a flashlight) to start a fire.

What's Needed
- Battery
- Thin wire or metal (from a twist-tie or staple, the spring from a ballpoint pen, steel wool, or a flashlight bulb filament)
- Tinder
- Kindling

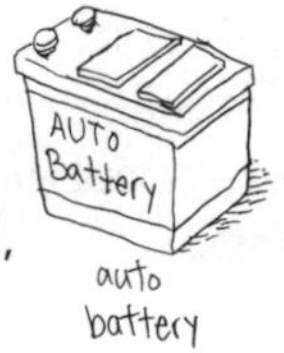

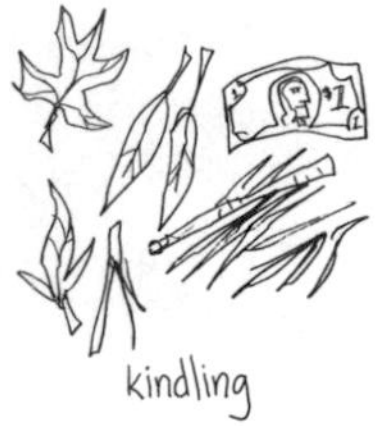

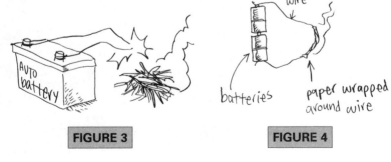

auto battery

kindling

What to Do
Attach two wires to the battery terminals. With tinder and kindling material at hand, use the other two ends of the wires to create a spark near the tinder (Figure 3). Once the materials starts to burn, add kindling material to keep the fire going. If you are using flashlight batteries, put them together, as in Figure 4.

You can also use thin wire strands or other small pieces of metal to start a fire by holding them across the battery (insulate your hands because the wire will get hot) and placing them in the tinder to allow it to burn.

wire

batteries

paper wrapped around wire

FIGURE 3

FIGURE 4

METHOD 4:
MAKE FIRE WITH A LENS

If it's bright and sunny outside, it's possible to use a lens to focus the heat of the sun on tinder material and start a fire.

What's Needed
- Lens (from eyeglasses—reading glasses only—a magnifying glass, binoculars, or telescope)
- Tinder
- Kindling

What to Do
With plenty of dry tinder available, aim the lens at it until it starts to smoke. Have other tinder material available to keep the fire going. When the tinder begins to burn, add kindling material.
See Figure 5.

FIGURE 5

METHOD 5:
MAKE FIRE WITH A REFLECTOR

What's Needed
- A reflector from a flashlight or car headlight
- Tinder
- Kindling

headlight
reflector

kindling

What to Do

You can use a light reflector from a flashlight or an old automobile headlight to focus the sun's rays on tinder material. If a headlight is to be used, carefully break away all the glass.

Position the tinder material in or in front of the reflector for maximum absorption of the sun's rays. With plenty of sunshine available overhead, and a little luck, the tinder material will get hot enough to catch fire. See Figure 6.

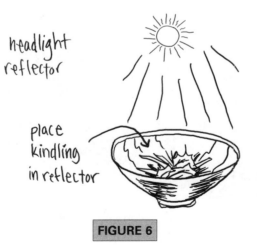

headlight
reflector

place
kindling
in reflector

FIGURE 6

METHOD 6:
MAKE FIRE WITH WATER

When positioned properly, water can act as a lens and focus enough of the sun's heat to ignite tinder.

What's Needed
- Water
- Jar or bottle
- Paper clip or twist-tie
- Tinder
- Kindling

What to Do
Pour about two teaspoons of water into a clear jar or bottle. Tilt the jar so the water rests in a corner at the bottom, position it

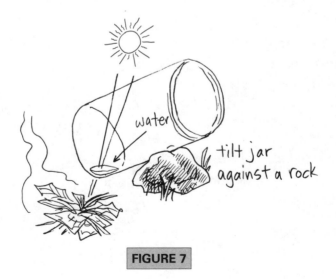

FIGURE 7

so the sun's rays shine through the water onto the tinder (see Figure 7), and ignite it.

Note: Other sneaky ways to make fire include using gunpowder from ammunition and chemicals from tablets. One of the most intriguing methods to be discovered is used by tribes in eastern Asia. They hollow out a bamboo cylinder and place small wood shavings at the bottom. When a long wooden rod is forced into the bamboo shaft (like a piston in a car engine), the rapid compression creates enough heat to ignite the wood shavings. The shavings are then poured on tinder to start a flame!

Rain Check:
Two Water-gathering Techniques

In a survival situation, finding water is crucial; without it, you can only survive a few days. Drinking water from the ocean can be dangerous because of its 4-percent salt content. It takes about two quarts of body fluid to rid the body of one quart of seawater. Therefore, by drinking from the ocean, you deplete your body's water, which can lead to death.

Fresh drinking water can be gathered from a variety of sources. This project will show how to gather rainwater and dew from the air.

COLLECTING DEW

What's Needed
- Clean towel or cloth
- Cup, bowl, or other container

What to Do
In the early morning, dew forms on grass, plants, rocks, and other large surfaces near the ground because these items have cooled and water vapor condenses on their surface. The dew can be easily gathered by laying a clean towel on the dew-covered area, dampening it, and wringing the towel out over a bowl. See Figure 1.

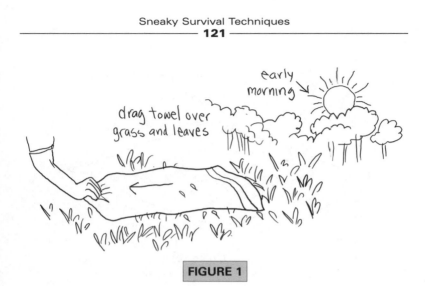

early morning

drag towel over grass and leaves

FIGURE 1

GATHERING RAINWATER

Rainwater, when available, is the preferred choice for drinking
because it does not require boiling or purification. It can easily
be collected by setting out items that you may already have.

What's Needed
- Cups, bowls, or other leakproof containers
- Plastic or vinyl material or a nylon jacket

What to Do
Place all available cups and containers where they can fill with
rainwater. If necessary, use waterproof material—plastic, vinyl,
or a waterproof article of clothing—as a substitute container,
as shown in Figure 2. Or make a container from a large leaf or
from coated paper, as shown in the bonus application Make
a Substitute Cup in the next setting.

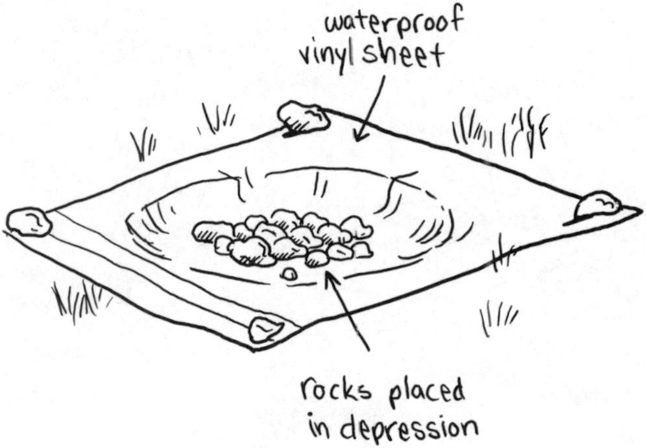

waterproof
vinyl sheet

rocks placed
in depression

FIGURE 2

Coming Extractions:
Get Drinking Water from Plants

Water is all around us in the air. The trick in obtaining it is to make it condense on the surface of an object and then collect it in a container.

EVAPORATOR STILL

An evaporator still can be made with a clear or translucent plastic bag and a large plant. It works by allowing the sun to shine through the bag and heat the plant, causing it to give off water vapor through its leaves. The water vapor condenses on the inside surface of the bag and drips down to the bottom. It can then be used for drinking water.

What's Needed
• Large plastic bag,
  preferably clear

What to Do
Gather green leaves or grass and place them in a plastic bag in a recessed area of the ground, as shown in Figure 1. Select an area where there will be plenty of sunlight. Or choose a plant or leafy tree branch, brush off any excess particles, wrap it in a plastic bag, and secure its opening with string, wire, or a tight knot; see Figure 2.

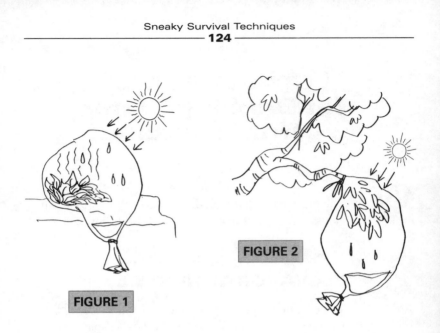

FIGURE 1

FIGURE 2

As the bag heats up, water from the leaves will evaporate and then condense in the bag as droplets that can be consumed later.

OVERGROUND SOLAR STILL

You can survive up to a month without food but only a few days without water. In the wilderness, there's always a concern about obtaining fresh drinking water. If you are near vegetation and have a large plastic bag available, you can quickly construct a solar still to acquire water.

A solar still uses heat to draw moisture from air, ground, or plants. It then collects the moisture droplets and condenses them into a container for drinking. Solar stills are easy to make, but the amount of water they produce will vary depending on their size, the amount of sunlight, and the terrain.

What's Needed
- Plastic bag, preferably clear, or plastic or vinyl material
- Cup or bowl or watertight container
- Rocks
- Stick
- Digging utensil

What to Do
The evaporator still proves that water from the air and from plant material can be trapped inside a plastic bag. With an overground solar still, you must dig out an oval or triangular trench and then another around it in an oval shape, as seen in Figure 1. Create the trench on an incline so that water will flow toward the end of the oval section.

First, place a tall stick in the center of the still and set the plant materials inside the center trench; see Figure 2. Next, cover

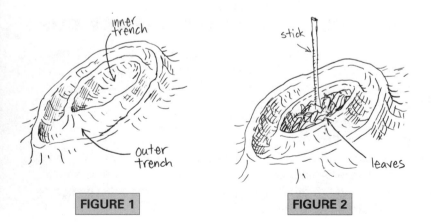

inner trench

outer trench

stick

leaves

FIGURE 1

FIGURE 2

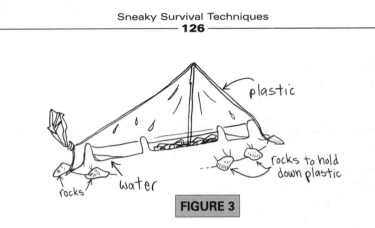

FIGURE 3

both the stick and both trenches with the plastic bag and hold it in place with rocks. Last, ensure that the bag end is closed and secure. Water from the plants will heat up in the sun, evaporate, condense on the inner surface of the plastic bag, and run down the sides and into the closed end of the bag in the outer oval trench, where it can be poured into a container later; see Figure 3.

BONUS APPLICATION:
H_2ORIGAMI—MAKE A SNEAKY CUP

Gathering or extracting condensed water from the air will be futile without a bowl or cup. You can make a sneaky cup from paper—preferably coated paper from a magazine—or from a large leaf from a tree.

What to Do
The following illustrations show a piece of paper with two sides. To clarify the folding technique, one side of the paper is shown as white and the other as gray. Each step is shown in its corresponding illustration figure number.

1. Start with a square piece of paper or large leaf.
2. Fold corner B diagonally on top of corner C.
3. Fold corner A, now point A, down as shown, forming a crease, and then unfold it.
4. Fold corner D, now point D, to the opposite edge, to the place where the first crease hits the edge of the paper.
5. Fold the paper on these two creases.
6. Fold the front (top) flap, corner B, down to cover all the layers. Fold the other flap, corner C, backward (there is only one layer to cover in the back).
7. Open the cup by pulling the front and the back apart.

The sneaky cup can be placed underneath a dripping condensation gathering area to save fresh water.

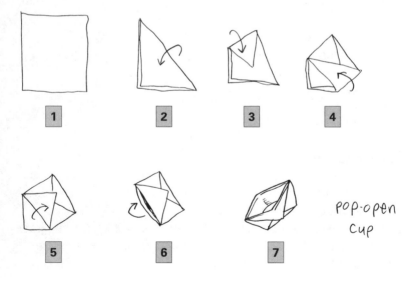

pop-open cup

Lens Crafter:
Build a Makeshift Telescope

If you're ever lost, it would be convenient, maybe even lifesaving, to be able to see long distances. Didn't bring your binoculars with you? Don't sweat it. Make a sneaky telescope from items you already have. Usually, a telescope requires two lenses. But there is a sneaky way to make a telescope with just one lens and aluminum foil.

What's Needed

- Toilet paper roll or stiff cardboard
- Lens (from a camera, eyeglass, or watch)
- Aluminum foil
- Rubber band
- Paper clip

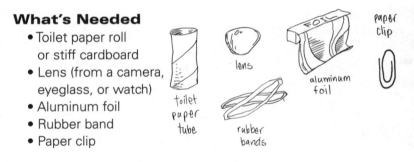

What to Do

The aluminum foil, from a snack bag or gum wrapper, will act as the telescope eyepiece lens. To make this work, you must poke a very tiny round hole in it. To make a good pinhole, stack up several layers of aluminum foil, poke the stack with a pin, separate the layers, and choose the one with the best small round hole. The size of the hole determines whether the images are sharp, blurry, or dim. Test different sizes until you obtain a happy medium.

Wrap the foil, with the pinhole in the center, around one end of the toilet paper roll and secure it with a rubber band, as shown in Figure 1. Cut a slit in the roll at its other end. Attach the paper clip to the lens and slide the lens in the roll by using the paper clip projecting through the slot as a handle; see Figure 2.

You can now use your pinhole telescope to create a zoom-lens effect by moving the lens toward the aluminum pinhole or away. Depending on the distance from pinhole to lens, the scene you see will be either upside down or right side up. It's very complicated to build a zoom-lens telescope with real eyepiece lenses, but if you use a pinhole it becomes simple.

FIGURE 2

tiny round hole

FIGURE 1

rubber band

slit

paperclip

lens

BONUS APPLICATION:
MAKE A SNEAKY MAGNIFYING GLASS

Need to make objects appear larger than they are? When you need to enlarge an image you can easily make a sneaky magnifier. Here are two ways to do it.

What's Needed

- Paper clip, twist-tie, or staple
- Clear window envelope or flashlight lens

paper clip or twist-tie

window envelope

foil end

lens end

FIGURE 3

What to Do

Method 1. Bend a paper clip into the loop shape shown in Figure 3 and dip the loop in water. As long as it is not too large, the water droplet will stay bonded to the paper-clip loop because of surface tension. Look into the water droplet, and the image you see will be magnified.

Method 2. Tear off the window part of the envelope and place a drop of water on it. Now as you look through the water droplet the image will be greatly enlarged; see Figure 4. The farther away you hold the envelope from the object, the larger it will appear.

In situations where you need to see up close, a sneaky magnifier will prove useful.

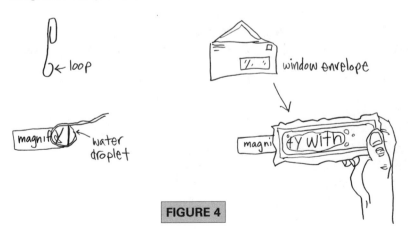

← loop

magnify ← water droplet

window envelope

magnify with

FIGURE 4

Smoke and Mirrors:
Sneaky Code Signaling

Being stranded in a remote area can fill you with fear. What's especially frustrating is seeing a plane or vehicle and not being able to get their attention to be rescued.

This chapter will supply various ways to signal for help. With a shiny object that reflects sunlight easily, you can signal to people and vehicles for assistance.

What's Needed
- Mirror, or belt buckle, metal pan or cup, or aluminum foil
- Reflective materials: canteen, watch, soda can, eyeglasses

What to Do
Using a mirror or other shiny object, point the light in one area and away in an SOS pattern (three long flashes, three short, and

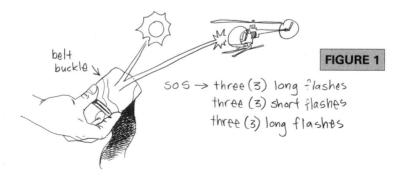

FIGURE 1

SOS → three (3) long flashes
three (3) short flashes
three (3) long flashes

three long). Repeat this sequence of flashes as long as possible while sunlight is available until rescued (see Figure 1).

The *U.S. Army Survival Manual* recommends:

1. Do not flash a signal mirror rapidly, because a pilot may mistake the flashes for enemy.
2. Do not direct the beam in the aircraft's cockpit for more than a few seconds as it may blind the pilot.

Haze, ground fog, and mirages may make it hard for a pilot to spot signals from a flashing object. If possible, therefore, get to the highest point in your area when signaling. If you can't determine the aircraft's location, flash your signal in the direction of the aircraft noise.

At night you can use a flashlight or a strobe light to send an SOS to an aircraft.

OTHER SNEAKY SIGNALING

When you're lost, use anything and everything as a marker to be seen by aircraft and search parties. Natural materials—snow, sand, rocks, vegetation—and clothing can be used as pointers to spell out distress signals. Follow the Ground-to-Air Emergency Code in laying out your markers.

SYMBOL	MESSAGE
I	Serious Injuries, Need Doctor
II	Need Medical Supplies
V	Require Assistance
F	Need Food and Water
LL	All Is Well
Y	Yes or Affirmative
N	No or Negative
X	Require Medical Assistance
→	Proceeding in This Direction

BODY SIGNAL	MESSAGE
Both arms raised with palms open	"I need help"
Lying on the ground with arms above head	"Urgent medical assistance needed"
Squatting with both arms pointing outward	"Land here"
One arm raised with palm open	"I do not need help"

See Figure 2 for illustrations of these signals.

To show that your signal has been received and understood, an aircraft pilot will rock the aircraft from side to side (in daylight or moonlight) or will make green flashes with the plane's signal lamp (at night). If your signal is received but *not* understood, the aircraft will make a complete circle (in daylight or moonlight) or will make red flashes with its signal lamp (at night).

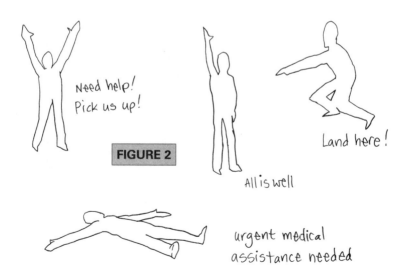

Need help!
Pick us up!

FIGURE 2

Land here!

All is well

urgent medical
assistance needed

Look on the Bright Side:
Make Sneaky Snow Glasses

In daytime, bright snow can blind you and cause damage to the eyes. What's needed is an emergency visor to filter out harmful reflected rays. You can easily make snow glasses out of a variety of things around you.

What's Needed
- Cardboard or leaf
- Material from clothing

What to Do
Cut slits in a piece of cardboard or other material and mount it on your head as shown in Figure 1. This will cut down on the ultraviolet light reflecting off the snow.

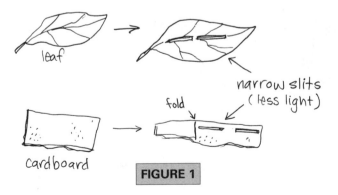

leaf

narrow slits
(less light)

fold

cardboard

FIGURE 1

Sneaky Snowshoes:
Walk on Top of the Snow

Very few of us are so prepared that we have snowshoes readily available in our house or car. In deep snow, your shoes or boots will sink and get wet. You'll waste energy and tire quickly, trying to lift your feet out of the snow to take the next steps. Also, you can't see what you are stepping on or how deep you may sink.

What's needed is a way to increase the footprint of the shoe—to spread your weight and allow more snow to provide increased support.

Snowshoes allow you to walk on the surface of the snow. If you're stranded without snowshows in the wilderness, you can quickly make a substitute set out of found items.

What's Needed
- Cardboard, or tree branches and plastic bags
- String, wire, or vine

Cardboard

plastic bags

string

What to Do
Cut or tear two pieces of heavy cardboard into the shape shown in Figure 1. Cut two holes in the cardboard to feed string, vine, or wire through and wrap it around your shoes and through your laces; see Figure 2.

Making sure that the bulk of the cardboard is in front of the shoe, bend the front upward, as shown in Figure 3. Do not tie

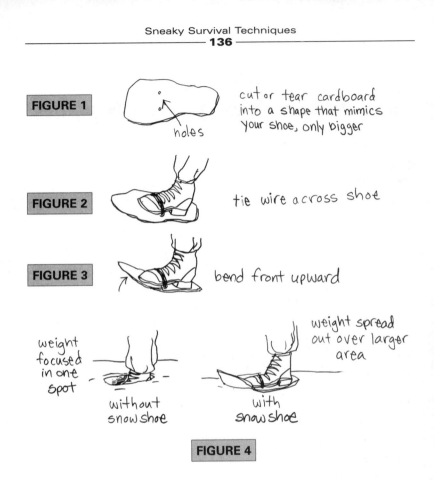

FIGURE 1 — cut or tear cardboard into a shape that mimics your shoe, only bigger — holes

FIGURE 2 — tie wire across shoe

FIGURE 3 — bend front upward

weight focused in one spot — without snow shoe

weight spread out over larger area — with snow shoe

FIGURE 4

the string too tightly. When walking in snow, your foot needs to tilt upward without gathering heavy snow.

If cardboard pieces or similar materials are not available, large tree branches encased in plastic bags can be used in a similar fashion. Now you can walk above the snow, as shown in Figure 4, and keep dry and safe.

Coldfinger:
Where There's a Chill, There's a Way

If you're like most people, you've imagined what it would be like to be stranded in a remote area in cold weather without winter clothes. If you are caught in a wintry climate without proper garments, you could get sick, injured, or worse. This project describes a sneaky way to use the things around you for comfort and safety.

What's Needed
- Leaves or paper

What to Do

If it's cold and you do not have the proper outer garments, you will rapidly lose body heat. Heat radiates and is lost from the body, especially from the head, neck, and hands. It is also lost via conduction when you perspire and touch other solid objects. Convection heat loss happens when you are directly exposed to cold winds.

You can reduce heat loss by using whatever is available to cover your head, neck, and hands. Use a plastic bag, newspaper, or undergarment to cover your head, including your mouth. (If the covering is plastic or paper, punch a hole to breathe through.) Covering the mouth as much as possible will direct some of the heat lost through breathing back toward your face as shown in Figure 1.

To create a pocket of air to trap body heat, make a sneaky cold-weather garment with leaves or paper. Assuming you are wearing a long-sleeved shirt and pants, gather leaves from trees or the ground and stuff them in your shirt, as shown in Figure 1. This will create a heat pocket to keep the air warmed by your body. Close sleeve and collar buttons securely.

Similarly, stuff your pants with leaves or paper, draw your socks over your pant legs or knot the ends to stop the leaves from falling out, and keep your hands in your pockets as much as possible; see Figure 2.

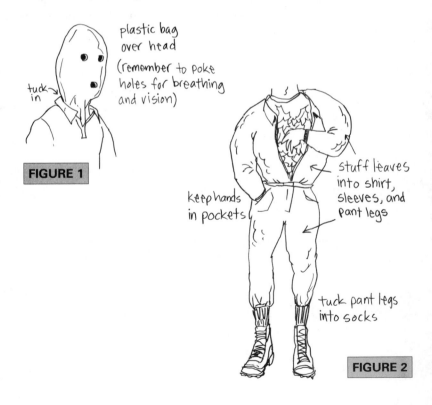

plastic bag over head (remember to poke holes for breathing and vision)

tuck in

FIGURE 1

keep hands in pockets

stuff leaves into shirt, sleeves, and pant legs

tuck pant legs into socks

FIGURE 2

Lost in Space?
Craft a Compass

If you're ever lost, you'll find a compass is a crucial tool. When markers or trails are nonexistent, a compass can keep you pointed in the right direction to get you back to a line of reference.

A compass indicates Earth's magnetic north and south poles. For a situation where you are stranded without a compass, this project describes three ways of making one with the things around you. For each method, you will need a needle (or twist-tie, staple, steel baling wire, or paper clip); a small bowl, cup, or other non-magnetic container; water; and a leaf or blade of grass. How simple is that?

METHOD 1

What's Needed
- Magnet—from a radio or car stereo speaker

What to Do
Take a small straight piece of metal (but do not use aluminum or yellow metals), such as a needle, twist-tie, staple, or paper clip, and stroke it in one direction with a small magnet. Stroke it at least fifty times, as shown in Figure 1. This will magnetize the needle so it will be attracted to Earth's north and south magnetic poles.

Fill a bowl or cup with water and place a small blade of grass or any small article that floats on the surface of the water. Place the needle on the blade of grass (see Figure 2) and watch

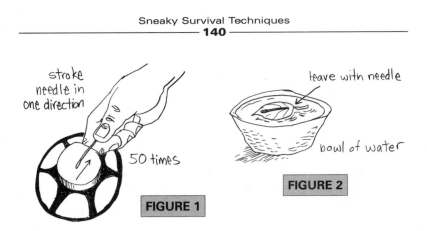

stroke needle in one direction

50 times

FIGURE 1

leave with needle

bowl of water

FIGURE 2

it eventually turn in one direction. Mark one end of the needle so that magnetic north is determined.

Note: To verify the north direction, see the next section, Road Scholar: Down-to-Earth Direction Finding.

METHOD 2

What's Needed

• Silk or synthetic fabric—
from a tie, scarf, or other garment

piece of silk or other synthetic fabric

What to Do

As in the first method, stroke a needle or paper clip in one direction with the silk material. This will create a static charge in the metal, but it will take many more strokes to magnetize it. Stroke at least 300 times, as shown in Figure 3. Once floated on a leaf in the bowl, the needle should be magnetized enough to be attracted to Earth's north and south magnetic poles. You may have to remagnetize the sneaky compass needle occasionally.

Note: You can also determine north and south with sneaky techniques in the "Road Scholar" chapter.

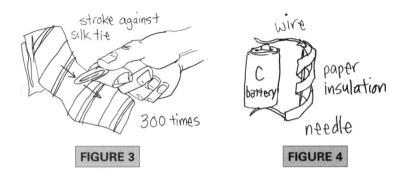

FIGURE 3

FIGURE 4

METHOD 3

What's Needed
• Battery

What to Do
When electricity flows through a wire, it creates a magnetic field. If a small piece of metal, like a staple, is placed in a coil of wire, it will become magnetized.

Wrap a small length of wire around a staple or paper clip and connect its ends to a battery, as shown in Figure 4. (For sneaky battery and wire sources ideas, see the "Industrious Light Magic" in Part III.) If the wire is not insulated, wrap the staple with paper or a leaf and then wrap the wire around it.

When you connect the wire to the battery in this manner, you are creating a short circuit—an electrical circuit with no current-draining load on it. This will cause the wire to heat quickly so only connect the wire ends to the battery for short four-second intervals. Perform this procedure fifteen times.

Place the staple on a floating item in a bowl of water, and it will eventually turn in one direction. Mark one end of the staple so that magnetic north is determined.

Road Scholar:
Down-to-Earth Direction Finding

If you're stranded without a magnetic compass, all is not lost. Even without a compass, there are numerous ways to find directions in desolate areas. Two methods are covered here.

METHOD 1:
USE A WATCH

What's Needed
- Standard analog watch
- Clear day where you can see the sun

Analog watch

What to Do
The sun always rises in the east and sets in the west. You can use this fact to find north and south with a standard nondigital watch.

If you are in the northern hemisphere (north of the equator), point the hour hand of the watch in the direction of the sun. Midway between the hour hand and 12 o'clock will be south. See Figure 1.

METHOD 2:
USE THE STARS

What's Needed
- A clear evening when stars can be viewed

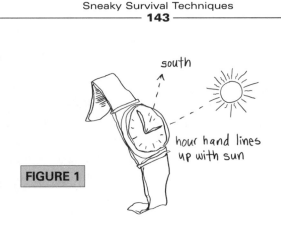

FIGURE 1

south

hour hand lines up with sun

What to Do

In the northern hemisphere, locate the Big Dipper constellation in the sky; see Figure 2. Follow the direction of the two stars that make up the front of the dipper to the North Star. (It is about four times the distance between the two stars that make up the

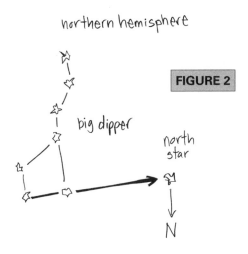

northern hemisphere

FIGURE 2

big dipper

north star

N

front of the dipper.) Then follow the path of the North Star down
to the ground. This direction is north.

In the southern hemisphere, locate the Southern Cross
constellation in the sky; see Figure 3. Also notice the two stars
below the Cross. Imagine two lines extending at right angles,
one from a point midway between the two stars and the other
from the Cross, to see where they intersect. Follow this path
down to the ground. This direction is due south.

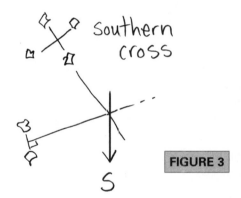

Southern cross

FIGURE 3

S

Harm and Hammer:
Devise Makeshift Weapons and Tools

Imagine being in a survival situation without any weapons, tools, or equipment except your knife. It could happen! You might even be without a knife. You would probably feel helpless, but with the proper knowledge and practice, you can improvise crude tools and weapons.

What's Needed
- Glass shards
- Thick plastic
- Rocks
- Tree branches
- Knife (if available)
- Strong vines, wire, strips of cloth, or shoelaces

glass shards

wire

What to Do
To make a club, locate a large rock with a slight indentation in the middle. If you cannot find one of this shape, you can fashion a groove in the rock by hammering the middle of the rock with another pointed rock or tool; see Figure 1. Next, find a strong stick or tree branch with a forked end and place the rock in the stick as shown in Figure 2. Last, secure the rock to the stick with the cross-wrap pattern shown in Figure 3. Test its strength by hammering an object and then check to see if it has loosened.

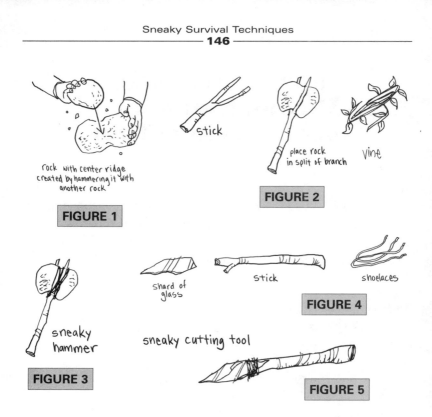

FIGURE 1

rock with center ridge created by hammering it with another rock

FIGURE 2

stick

place rock in split of branch

vine

FIGURE 3

sneaky hammer

FIGURE 4

shard of glass

stick

shoelaces

FIGURE 5

sneaky cutting tool

To make a knife, locate a sharp shard of glass or stone and a stick, as shown in Figure 4. The stick should have the short end of a branch still present so the glass or stone can be braced against it. Wrap the glass to the stick securely, and test it for strength and excess movement; see Figure 5.

Note: Sneaky weapons and tools must be used cautiously for safety reasons.

Pocket Protectors:
Sneaky Tools and Survival Kits

How many times have you fumbled in the dark and wished you had a flashlight? Or needed a screwdriver or pliers? Many everyday situations and emergency conditions can be resolved with the proper tools, but most people do not carry a bulky toolbox with them at all times.

This project illustrates a variety of innovative and compact multitools and survival supplies you can carry around in your pocket. For trips away from urban areas, you can assemble a sneaky survival kit that can be carried inside a pen!

MULTITOOLS AND MORE

A new category of ultralight multitools is available that should be carried all the time. These novel pocket-sized items should become your everyday accessories.

To be prepared, you should always carry these items in your pocket or on your key ring: Figure 1 illustrates the usefulness of this collection.

Multitool (includes four screwdrivers, pliers, cutters, and wire benders)
Mini flashlight
Credit card multitool (includes a compass, magnifying glass, and serrated blade)
Plastic whistle

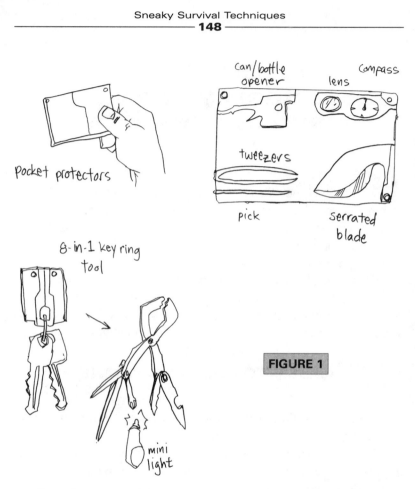

pocket protectors

can/bottle opener

lens

compass

tweezers

pick

serrated blade

8-in-1 key ring tool

mini light

FIGURE 1

For a low-cost mini-portable motion alarm, obtain an XP-4 Spy Pen from Wild Planet (see www.wildplanet.com). Sold in toy and discount stores for under $15 at the time of this writing, the gadget-filled pen includes a magnifying glass, telescope, flashlight, and a real working motion detector.

For information concerning multitools and gadgets, check the following company Web sites:

http://www.advanced-intelligence.com
http://www.beprepared.com
http://www.berberblades.com
http://www.colibri.com
http://www.equalizers1.com
http://www.leatherman.com
http://www.spyderco.com
http://www.swissarmy.com
http://www.swisstechtools.com
http://www.toollogic.com
http://www.topeak.com
http://www.wildplanet.com

SNEAKY SURVIVAL KITS

Why not prepare for the worst when you're traveling? Most people wouldn't want to carry a backpack full of equipment, but if there is a way always to have available the minimal items for a survival situation in a package as small as a mint box or a pen, who will argue against that?

Assuming you already have a multitool, mini flashlight, credit card multitool, and plastic whistle on your key ring, here are a few other items you should take with you when you travel outside of urban areas.

Magnetized needles
Strong nylon thread
Small safety pins (for fishing hooks)
Thin wire
2 small watch batteries, 1½ volts each
Small roll of duct tape
Small roll of aluminum foil (as a signal mirror)

Dental floss
Aspirin
Tiny multivitamins
Tiny candle
5 match heads (placed end to end and wrapped in foil)
$20 bill
Sugar

Believe it or not, you can pack all these items inside a hollowed-out mint container, cigar holder, or fat writing pen! See Figure 1. (When using a pen, remove all of the inner parts and seal both ends.) With these compact tools and supplies, you will greatly improve your odds in survival situations. For more information about assembling survival kits, see the Resources section that follows.

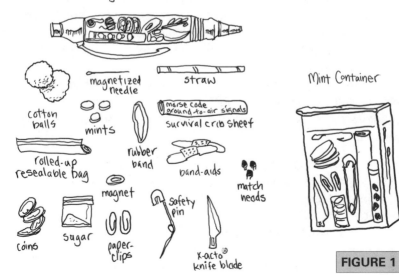

Big-wide Pen

magnetized needle · straw · Mint Container

cotton balls · mints · morse code ground-to-air signals survival crib sheet

rolled-up resealable bag · rubber band · band-aids · match heads

magnet

coins · sugar · paper-clips · safety pin · X-acto® knife blade

FIGURE 1

RESOURCES

You can always do more with what's around you. Indeed, one day it may be necessary for you to improvise to survive. Look around you and think about the many ways you can put other items to alternative use. What can you make with a radio control car, walkie-talkie, windup toys, retractable badge holders, toy dart guns, toy robots, or rubber-band-powered items? Plenty.

Think of the many ways a simple item like a flashlight can be used. Ponder its every part. Here are just a few sneaky functions:

The case can be used to collect water, to block a window from opening, as a safe for small valuables, as a fire extinguisher container, and as a weapon.

The reflector can be used to start a fire or to signal for help.

The bulb's shell can cut small items.

The batteries can be used to start a fire or to magnetize a piece of metal.

The lens cover can be used with water as a magnifier or to start a fire.

The bulb filament can be used to start a fire.

What other parts of the flashlight have not been mentioned? What can they be used for? Your way of looking at things is your greatest resource. To borrow a business phrase, "Think outside the box." The projects in this book should have expanded your thoughts about what you can do with everyday things.

Note: Be sure to visit **Sneakyuses.com** for extra project information, Web links, and resourceful contest information and to post your imaginative discoveries.

USEFUL WEB SITES

Frugal and Thrift Sites

choose2reuse.org
freegiftclub.net
frugalcorner.com
frugalitynetwork.com
getfrugal.com
make-stuff.com
ready-made.com
Recycle.net
thefrugalshopper.com
thriftydeluxe.com
wackyuses.com
watchthepennies.com

Pocket Tools and Gadgets

advanced-intelligence.com
berberblades.com
casio.com
colibri.com
equalizers1.com
girltech.com
johnson-Smith.com
leatherman.com
netgadget.net
nutsandvolts.com
robotstore.com
scientificsonline.com
spyderco.com
spy-gear.net
swissarmy.com

swisstechtools.com
topeak.com
undercovergirl.com

Survival
backwoodsmanmag.com
basegear.com
beprepared.com
campmor.com
equipped.com/
fieldandstream.com
productsforanywhere.com
ruhooked.com
selfreliance.net
simply-survival.com
skillsofsurvival.com
Survival.com
survival-center.com
Survivaliq.com
Survivalx.com
wildernesssurvival.com

Home Security Sites
mcgruff.org
ncpc.org
safesolutionsystems.com
X10.com®
youdoitsecurity.com

Science and Technology
about.com
boydhouse.com/crystalradio

Craftsitedirectory.com
discover.com
hallscience.com
HomeAutomationMag.com
howstuffworks.com
midnightscience.com
RadioShack.com
scienceproject.com
Scientificsonline.com
scitoys.com
thinkgeek.com
wildplanet.com

Other Web Sites of Interest

Almanac.com
craftsitedirectory.com
Doityourself.com
Movie-Mistakes.com
Nitpickers.com
Oopsmovies.com
Popsci.com
Popularmechanics.com
rube-goldberg.com
Smarthome.com
tbotech.comrotorsportz.com
Thefunplace.com
Tipking.com

RECOMMENDED READING

Books

David Borgenicht and Joe Borgenicht, *The Action Hero's Handbook* (Quirk Books)

Robert Young Decton, *Come Back Alive* (Doubleday)

Ira Flatow, *They All Laughed . . . From Light Bulbs to Lasers: The Fascinating Stories Behind the Great Inventions That Have Changed Our Lives* (Harper Perennial)

Joey Green, *Clean It! Fix It! Eat It!: Easy Ways to Solve Everyday Problems with Brand-Name Products You've Already Got Around the House* (Prentice Hall)

———, *Clean Your Clothes with Cheez Wiz: And Hundreds of Offbeat Uses for Dozens More Brand-Name Products* (Prentice Hall)

———, *Joey Green's Encyclopedia of Offbeat Uses for Brand-Name Products* (Prentice Hall)

Lois H. Gresh and Robert Weinberg, *The Science of Superheroes* (John Wiley & Sons)

William Gurstelle, *Backyard Ballistics* (Chicago Review Press)

Garth Hattingh, *The Outdoor Survival Handbook* (New Holland Publishers)

Dave Hrynkiw and Mark W. Tilden, *JunkBots, Bugbots, and Bots on Wheels: Building Simple Robots With BEAM Technology* (McGraw-Hill Osborne Media)

Vicky Lansky, *Another Use for 101 Common Household Items* (Book Peddlers)

———. *Baking Soda: Over 500 Fabulous, Fun, and Frugal Uses* (Book Peddlers)

———, *Don't Throw That Out: A Pennywise Parent's Guide* (Book Peddlers)

————, *Transparent Tape: Over 350 Super, Simple, and Surprising Uses* (Book Peddlers)

Joel Levy, *Really Useful: The Origins of Everyday Things* (Firefly Books)

Hugh McManners, *The Complete Wilderness Training Book* (Dorling Kindersley)

Forrest M. Mims III, *Circuits and Projects* (Radio Shack)

————, *Science and Communications Circuits and Projects* (Radio Shack)

Steven W. Moje, *Paper Clip Science* (Sterling Publishing Co.)

Bob Newman, *Wilderness Wayfinding: How to Survive in the Wilderness as You Travel* (Paladin Press)

Tim Nyberg and Jim Berg, *The Duct Tape Book* (Workman Publishing Company, 1994)

————, *Duct Tape Book Two: Real Stories* (1995)

————, *The Ultimate Duct Tape Book* (1998)

Larry Dean Olsen, *Outdoor Survival Skills* (Chicago Review Press)

Joshua Piven and David Borgenicht, *The Worst Case Scenario Survival* (Chronicle Books)

————, *The Worst Case Scenario Travel* (Chronicle Books)

Royston M. Roberts, *Serendipity* (John Wiley & Sons)

US Air Force Search and Rescue Handbook (Lyons Press)

US Army Survival Handbook (Lyons Press)

Jim Wilkinson and Neil A. Downie, *Vacuum Bazookas, Electric Rainbow Jelly, and 27 Other Saturday Science Projects* (Princeton University Press)

John Wiseman, *The SAS Survival Handbook* (Harvill Books)

Magazines

Backpacker
E Magazine
Mother Earth News
Nuts and Volts
Outdoor Life
Outside
Poptronics
Popular Mechanics
Popular Science
Ready Made
Self-Defense